Also by Bert Tucker:

Blind Quest: Deceived by Experience
Blind Quest: Avoiding Pitfalls

LIFE FORCE
MARS

CREATING A NEW HOME FOR MANKIND

BERT TUCKER

iUniverse, Inc.
Bloomington

Life Force Mars
Creating a New Home for Mankind

Copyright © 2011 Bert Tucker

iUniverse books may be ordered through booksellers or by contacting:

iUniverse
1663 Liberty Drive
Bloomington, IN 47403
www.iuniverse.com
1-800-Authors (1-800-288-4677)

ISBN: 978-1-4620-1247-3 (pbk)
ISBN: 978-1-4620-1248-0 (ebk)

Printed in the United States of America

iUniverse rev. date: 5/18/2011

For Rob, Tim, and Billy, who could see this happen.

CONTENTS

The substance of this work owes much to assistance provided by others.

Col. (Ret) Don Peterson—Air Force fighter pilot, NASA astronaut, Shuttle Crew (for extensive space and robotic technical advice and their human implications)

Lynn Harper, NASA Ames Research Center (for expert state-of-the-art horticultural technical advice, insights that saw beyond the veil, encouragement when it was most dear). Many others at NASA ARC have been very helpful with e-mails and by telephone.

LTC (Ret) Sam Gates—Military Engineer (for technical advice on nuclear power and critical review)

I owe many thanks to Nathalie Cabrol and Carol Stoker of NASA Ames RC, Roy Christini and Robin Oder of ALCOA, Roger Clark and Jennifer Blue of USGS for early guidance.

Much in the way of special thanks to JPL and Malin for orbital photographs including those they provide on the Internet. Much in the way of description of geological features is derived and often paraphrased from the maps, photo maps, and geological maps with text produced by the US Geological Survey for NASA and available to the public. Individuals at USGS have helped by e-mail and telephone with special thanks regarding place names.

Toward that end, I have used numerous examples of how this could happen. One example is the use of in-hand small construction equipment.

President Barack Obama has cut NASA's manned space exploration program, effectively terminating the serious manned exploration of the solar system. However, existing rocket propulsion systems are substantial and may be enough with supplemental Mars landing modules and the like.

This story involves substantial common technology that is referenced, but it includes no historical characters or incidents requiring references. There have been no missions such as I mention in my story, which were intended to explore for a suitable manned landing site on Mars.

However, I have been working on this manuscript so long that I found newly developing real science kept stepping on my fictional technology already drafted into my story. I find it reassuring that I am staying in the bounds of what could really be developed.

Lost Ambition

Notice the subtitle. When we go to Mars, we will be establishing a new home for humanity! That is the entire thrust of this story. This is not just a real science fictional story—it's an illustration of how we could, in an unorthodox way, actually begin sending humans to live on another planet today.

Forty-seven years ago, I took a comprehensive, non-credit course as a graduate student in physics at Louisiana State University. That course was presented by the NASA technical team then designing and developing the Apollo program. That mission required vision and commitment to send men to the moon. The people that developed the Apollo missions were imagining something that had never been done before. My course of study included very substantial documentation of the various aspects of their project and gave me the privilege of imagining along with them.

I'm bouncing off of the Apollo experience in writing this story—and the planned unmanned Mars exploratory missions—although my approach is not at all the approach proposed by NASA. My approach is thinking outside of the box and would involve a high degree of risk.

Events in this story can move far faster than a normal story because the resources available to make progress are not just the people going to Mars. In particular, there are all the resources of the JPL contingent and at NASA actively directing AI robots and AI robotic equipment on Mars all of the time, day and night. That means the real limitation is the humans working from Earth, how many robots and how much robotic equipment is available on Mars at that time, not just the few humans who are actively on Mars.

My story is based upon today's real world. There are discoveries in this story, but they are realistic in the sense that most minerals found on Earth are likely to exist somewhere on Mars. So this story is somewhat like the Apollo experience: realistic. How can we colonize Mars with current technology, or at least technology that lies on the horizon? We have never had an Apollo sequel, even to the moon. And we have never brought anything back from another planet. We don't yet have a plasma rocket that may eventually cut the travel time to Mars, but NASA did pursue a plasma rocket project.

We live in a time when the United States is no longer the huge technical or political power of its day. We live in a time of terrorism, overpopulation, limited fossil fuels, overextended budgets, exhaustion of natural resources, and global warming.

Perhaps we are not seeing the obvious. We keep mentioning global warming, but we are not seeing the obvious major changes in our weather already evident today. At this writing in early February 2011, we have just experienced the largest continent-covering extreme winter storm in perhaps a century. Why now?

Under these conditions and our current deep economic recession, we know that we must be frugal, we must use robotics extensively under guidance from Earth, and we must make the Mars operation self-sufficient as soon as possible. They all play a part in my story.

We know a lot about Mars; we have gained much perspective from the remarkable unmanned explorations in recent years.

This story uses real Mars place names, real Mars topography, and the real Mars environment.

For example, a NASA *Ares I* rocket was successfully launched on October 28, 2009 for a two-minute powered flight. The 327-foot-tall *Ares I-X* test vehicle produced 2.6 million pounds of thrust to accelerate the rocket to nearly 3 g's and Mach 4.76, just shy of hypersonic speed, rising to a suborbital altitude of 150,000 feet after the separation of its first stage, a four-segment solid-rocket booster. There was a simulated upper stage and Orion crew module. The Orion crew module was planned to complete the *Ares I* Crew Launch Vehicle (CLV).

Something like the *Ares I* CLV is anticipated in this story as the vehicle that transports the crew module (the colonial crew and equipment) to Mars. This base development module is referred to as the BDM.

At the beginning of this story, day-to-day human progress on Earth had come at a price. Rich natural resources were largely depleted and competition for natural resources was becoming intense. Rising prices were cutting into the affluent lifestyle of the most powerful nations. In other words, the situation is *today*.

So why should Mars interest us? It offers a laboratory for survival without fossil fuels. It also offers a minerals-rich surface with more land area than Earth. However, it is far more distant than the Moon, bitterly cold, completely arid on the surface, and its atmosphere is almost non-existent. Transport of anything to Mars is extremely risky, very long in transit, and formidably expensive.

Mars does offer the prospect of one reliable native energy resource—radioactive uranium and derivable plutonium. Hence Mars has independently derivable nuclear power. Initial enriched fuel, equal in energy content to kilotons of fossil fuels, could be economically shipped from Earth until native uranium ore could be discovered, and uranium metal refined from the ore.

The native uranium metal could be enriched to 5 percent in the radioactive isotope, which provides fuel or electrical power generation.

Despite the March 2011 9.0 earthquake and subsequent tsunami in Japan, this is very viable, particularly since there are no earthquakes or tsunamis on Mars.

So exploring Mars and developing a human base there is much more than scientific discovery. It could mean the survival of humanity.

The situation on Earth continues to worsen with the depletion of natural resources, the increase of global warming, continuing increase in all types of global pollution, no end in sight to population explosion, and of course the ever-present terrorist attacks threatening nuclear devastation. We would do well to secure a foothold on Mars—the only prospective living space for humanity and Earth's precious biological family.

Without ambition there is no hope. *Life Force Mars* is meant to realistically inspire ambition.

There is a human story here. This is not just adventure; it is a scientist-engineer's road map with some excitement thrown in for fun.

PRECOGNITION: WHERE DOES THIS STORY TAKE US?

The time is 2016 when the colonists are arriving near Mars.

Every spring, the warming effect of the sun gradually melted the dry ice polar cap on Mars, releasing a powerful spiral of carbon dioxide. The carbon dioxide cloud grew into a dynamic cyclone of air across the seabed of Mars's once-huge northern ocean. It picked up the pervasive dust that covered the entire surface of the dead planet, creating a thick blanket that obscured the surface, blocking the weak sunlight and obstructing the passage of radio waves. The Earth Control team used radio communications from Earth to the Marscom geostationary satellite to Mars Control within the wall of the Genesis Crater to direct affairs on Mars.

Artificial Intelligence robots had begun the daunting chore of carving arched passageways and domed rooms to begin a base for people and provide electrical energy to an ambitious artificial life-support environment. NASA was releasing the aggressive force of life and was trying to create a new home for humanity in the process.

But in this instance an enormous dust storm was not expected and the habitat was just barely begun to provide living space for the first colonists. Six months earlier in 2016, four powerful *Ares* rockets had blasted off from Cape Kennedy and were approaching a home that was not ready for two of the four human colonists. They were trapped. They had nowhere else they could go. They were committed.

Four ships carrying the astro-colonists were approaching Mars. They had been launched from Earth during the latest Mars opposition. They were staged ten days apart. Mars surface-radio contact had been lost by the lead ship, *Venture 1,* four weeks earlier. *Venture 1* was carrying Tim, an astronaut, and Joan, a planetary geologist. Two nuclear-powered electrical generators installed within the wall of a meteor impact crater were barely limping along. Whether the generators could continue production of power until the landing was in doubt. Without power, all habitat operations and life support for plants and animals on Mars would die. The obscuring dust cloud in any event prevented radio-transmitted ground guidance systems from directing *Venture 1* to a safe landing.

Venture 2, carrying Rob, the habitat's engineer, and Mary, the habitat's biologist, was only ten days from Mars when a small meteorite struck with the impact momentum of over 30,000 miles per hour velocity, puncturing both sides of the hull. The drop in air pressure immediately caused atmospheric air to spew outward into space through the small holes, perhaps leaving the astronauts without enough oxygen until they were due to land.

Rob's response was reflexive. "What the hell! We just lost our life-support air, Mary! Seal your jumpsuit *now*!"

The jumpsuits were designed for light-duty pressurization. A pressure suit helmet sealed quickly into place and snap air connections quickly inflated the suits, bringing them up to normal pressure. Alarms spread to distant Earth and *Venture 1* at the speed of light. Trace dyes automatically followed the flow of air to the hull ruptures.

Rob grabbed a patch and was locating the first rupture at the entry point. Joan had another patch and was applying her patch to the exit hole. The sturdy fiberglass-reinforced, self-adhesive patch was slapped into place, and then a second, larger patch was carefully applied over the first.

Venture 1 had been approaching Mars for landing when they received the alarm from *Venture 2*. They all knew hull ruptures were usually catastrophic.

The two couples were the only people going to Mars onboard the *Ares* rocket *Venture* fleet. They had worked together for five years. They were intimately bonded. Tim was their leader. He immediately squeezed his radio transmit button. "V-2, report immediately."

The response was immediate. "We're still here, but we lost a lot of air. Cabin pressure is very low, but our suits are holding up okay. We'll have to land in our current light-duty pressure suits. Our computer is telling us that we're going to be dangerously close to running out of oxygen before we land."

Tim replied, "We'll be going down in just two hours. Give me your latest sit-rep just before we descend. We'll attempt to guide you down if we get any radio link through the dust cloud as you land."

Mary knew everything from here on in was critical. "We're setting up to use emergency oxygen bottles. The module pressure is not enough to support us normally. We'll be out of contact until we descend. We're counting on you. We love you both."

Joan was extremely concerned. "Listen you two, we really need you and love you. If there is any way to fix things, we'll do it. This is a rotten turn of events."

Tim continued, "Stay in contact with Earth Control. They may be able to stretch your oxygen supply."

He continued his preparations for landing until the computers indicated that they were in position for descent.

Rob gave his last report to Tim. "Grip hands. Remember *Apollo 13* landed safely despite its many equipment failures."

Retro rockets fired and *Venture 1* began its descent. The crew module assumed its planned "orbit-to-surface" trajectory, targeting their landing zone near Craterwall, their new home.

CHAPTER 1:

RADICAL PLAN

THE STORY BEGAN in the fall of 2011.

Gus Hoover had never been known as a radical at NASA, but that was about to change. "I'm telling you," he said as he leaned forward over the conference table, "you can just leave them there. The astronauts don't have to come home from Mars!"

Gus was a lanky Texan and he was Texas grim. He had also been the director of manned space flight missions under the shuttle program. "We're faced with an impossible prospect. Money to fund a full-scale human mission just is not available ... so we don't do that. We send them to Mars, but we don't bring them back ... at least not right away. With that condition and a bit of luck, we can send people to Mars *now*."

Dave Bagnal was the NASA chief executive, trim and tailored through and through. He slid back in his chair as though to escape Gus and looked over toward Louise Kruger. She was tall, slim, and blonde. She was their chief technical expert. Her eyes narrowed.

Gus was earnest. "We build a base with only essentials using artificial intelligence robotics controlled from Earth. Then we send people to live in that base and expand it. We've sent many missions to Mars, but we have no experience in bringing anything back, so we don't."

Dave frowned. Louise was modestly attractive, technically astute, and intent. Her body language agreed with Dave. Gus continued.

"Our mission concept has always been to send astronaut explorers to Mars at one Earth-Mars opposition, when Earth and Mars are closest in space, and bring them back at the next. That requires supporting them with supplies and life support for two and a half years and sending a lot of equipment and fuel to get them back … also a high-risk business.

"That approach is so much pie in the sky. It is unreasonable! It's impractical as an engineering project and the costs would be so extreme they would never be funded. I've been considering this problem for years and I finally realized the obvious. We can't afford to fund and develop the return from Mars now. We must cut our Gordian Knot. The astronauts must stay on Mars as colonists indefinitely. They wouldn't come right back."

He looked at his boss, wondering if Dave would buy into his seemingly crazy plan. "With the Ares rocket, we actually have the means within our grasp to begin a colony on Mars now."

Dave and Louise smiled grimly. Dave was Gus's boss in charge of all NASA exploration projects. He issued his challenge. "So you think you can make this happen?"

"What I'm suggesting is to increment the mission in manageable parts. I'll give you a stepwise approach that takes off from the present unmanned missions and one much more likely of success. We'll sneak up to the threshold step by step. Each step will seem like just more exploration until ultimately the choice of sending people to Mars will seem obvious. A team will be working on an AI-robotics-controlled project to build a working base populated with animals and plants within a life-support system. Once we have that kind of base on Mars, who would deny the team that created that base the opportunity to go live in that base. They'll go to Mars and they will stay on Mars!"

Louise managed planetary projects. She shook her head. "Building that kind of base would be a huge project by itself."

Gus was extremely earnest. "First, we'll use resources native to Mars to make this base virtually self-sufficient. The primary exception is nuclear fuel for a power plant. But from the outset that's relatively easy to transport.

"We've a good idea where we can find water… up near the fossae complex on the northeast edge of the Tharsis Bulge. Fossae are rare, but tend to occur in clusters. Fossae were probably formed by the melting of subterranean ice along a fissure or fault line many eons in Mars's distant past. The existence of the fossae is strong evidence water was there ages ago when Mars was warm. Fissures developed as the crust cooled. That crack in the crust let early Mars core heat into the aquifer which then evaporated its water into the air, much like the fumaroles in Yellowstone Park.

Gus continued. "Those huge fossae are really just sinks with no outlet. They're close to a thousand feet deep which could mean there's an aquifer hundreds of feet thick just off to the sides of the depressions. When the water was gone the ground collapsed leaving those boxed in canyons. Water was there once and I bet there will be water nearby to those fossae today."

Dave was not easy to sell. "How do you build this base? Mars has no building materials or reinforcements if you choose to go underground."

Gus laid some pitch black stones on the table. The others picked them up, discovering they were heavy and sturdy.

"I haven't seen those on the surface, particularly not shaped like these as building blocks," said Louise.

Gus responded. "Those were fused in molds using a thermite chemical reaction. The raw material we used has a chemical content like Mars surface dust, which we magnetically enriched in iron oxide. We then added aluminum powder so the mixture could produce a highly exothermic chemical reaction when heated. If ancient Romans could build with stones, then so can we, and these are very sturdy. We bore into the wall of a relatively small impact crater, carving out passageways and rooms. We line those spaces with domed and arched ceilings of Marscrete, my name for what you are holding. We need the ballast weight of the crater on the ceilings to contain the air pressure of a life-support atmosphere. Constructing within a raised crater wall means we don't really go underground."

Dave seemed to be wavering. "How large should the crater be—not that there's any shortage of craters."

"I've been to the lip of Meteor Crater just east of Flagstaff, Arizona. That crater would do just fine—say about a kilometer across, a little over half a mile. That crater right here on Earth would provide a perfect test site for our concepts.

"Once we retrieve proof of water from a couple of hundred meters deep drilling, we will have good reason to set up exploration in that area for native minerals. As you know, surveys show that there is water ice at about 45 degrees north, which is where we would be testing. And we already have surface dust everywhere that can be concentrated magnetically, even into rich iron ore. The composition of much of the surface of Mars is not all that different from here on Earth. We've known that ever since the *Viking* missions in 1976."

Dave knew that Gus could be tenacious. "So give me a conceptual plan. I tend to agree with Louise. This is bigger than you think."

Gus said, "Just for planning, I need four specialists, and I have people in mind. First, an astronaut with spacecraft experience to lead the design of the transport to Mars and to design a life-support system for the base. Second, a field engineer to design and oversee the building of the base and its outside facilities. Third, a planetary geologist familiar with Mars to find the native materials we will need. Fourth, a life sciences, astro-biology expert to work with the engineer in creating a living space for people, plants, and animals. Three of them already work for NASA. The engineer is military—accustomed to field construction under extreme conditions. All of them are innovative and proven at overcoming severe obstacles."

The next step was to find people with the professional background and vision to demonstrate how a manned exploration of Mars might be accomplished. Then they needed to develop a project proposal that could realistically ignite the fire of manned interplanetary exploration.

Gus was not just looking for expertise. He wanted a strong, cohesive team and a leader for his group that was daring, but not one to overlook risk. Major Tim Randall, US Air Force, was a youthful astronaut just coming off of a tour on the ISS, the International Space Station. He was part American Indian, a.k.a. Native American, held a PhD in astrophysics, and was very familiar with the technology involved. Randall had demonstrated a lot of initiative working with advanced AI robots and was not frightened of the emerging technology.

Gus also wanted someone who was a leader—someone who could lead people across more than a hundred million kilometers of space to build a human base on Mars. Randall was capable of developing the space transport part of the plan and could remotely fly the gossamer-light, helicopter-like structure that was their aerial observation platform. He knew what it was like to live in space. Mars—with 3/8 Earth's gravity and 1 percent of Earth's atmospheric pressure—would not be entirely foreign to him.

Major Robert Anderson, US Army, was a proven military engineer who had constructed roads, airfields, and a wide variety of structures in Afghanistan and Iraq. He had worked miracles in the war against terrorists on their own turf. He was a strong team player with medals to commend his bravery. His PhD was in civil engineering, but he really stood out with his remarkable innovations. Gus wanted a strong partner for Randall—one who had a reputation for not just solving problems, but carrying his projects to a higher level.

Dr. Mary Hoffmann was a proven quantity at the NASA Ames Research Center, where she had developed varieties of grain, vegetables, and fruit that would thrive in hydroponics and ultimately in engineered soils. She had bioengineered new varieties of many plants with unique characteristics. She was widely respected—not only on the Ames campus, but also in the academic and entrepreneurial scientific community.

Dr. Joan Wall was their geological ace. She had worked for NASA for only four years, but had performed the analyses on the minerals discovered by the biannual explorations of Mars. Gus saw her as an inside and outside explorer and top analyst. She was part Mexican Hispanic, held her PhD in astrogeology, and was on loan at the moment to Cal Tech's Jet Propulsion Laboratory, JPL, working on the 2012 projects. She had cut her teeth on the real projects now exploring Mars. She had found new deposits that others overlooked. She was a strong proponent of deep core drilling and was determined to be the first to discover substantial amounts of subterranean water on Mars.

All four were innovators on the frontier. They were all outgoing, yet had a knack for developing strong interpersonal relationships. All had credentials and accomplishments that were the envy of their colleagues.

These elements were essential if the project was to succeed. All the technical expertise would mean nothing without social cohesion. The challenge for Gus and for the four team members was how to build long-term social cohesion—and then find a way to hold the group together long enough for it to produce the desired effect. And only Gus knew that was his goal.

The three NASA people had no glimmer of what was coming. They accepted the opportunity to explore interplanetary issues at Houston with a select group of colleagues. They knew Dr. Gus Hoover by reputation. His invitation was an opportunity that they would not miss.

For Rob, it was another matter. He received orders to report to NASA on an extended, but unspecified assignment. He knew his expertise and did not fathom where it could be used for space purposes. He was a little surprised to find out that he would be staying in a long-term-occupancy mini-suite.

Mary and Joan discovered each other at the reception counter of the residence hotel where they began comparing notes. Rob, upon discovering two attractive women, casually overheard their conversation and introduced himself. Since they were to attend the same session, speculation arose. He was not disappointed at all to discover that both women were single, but he could not find any clue except that Mary was working on the Mars explorations.

They were preparing to move into the dining room when Tim showed up. He lived nearby and Louise had let him know where the others would be staying. His confident astronaut demeanor was most apparent. Rob was torn between enjoying spirited conversation with the ladies and Tim's astronaut background until he discovered they were of a single mind. What was afoot? The two military men quickly hit it off. The four singles were soon sizing up each other and anticipating their future, particularly when they discovered that they were all there with the same invitation. A tantalizing adventure awaited them.

CHAPTER 2:
ALTERNATE VISION

THE PROSPECT OF developing the project depended upon technical concepts that were already normal—even in everyday construction—on Earth. Still they needed to be clearly understood if the project was to be considered worthy. What followed was the thrashing out of those concepts. Everyone had to agree upon and understand them.

Gus and Louise met with their Mars Base team the next morning. Louise was in charge of their work and addressed them.

"You were selected to develop a conceptual plan and to find a way to construct a working habitat on Mars. This is a giant leap forward from our 'here and there' explorations to date. We're considering a real working base we can seed from Earth, but constructed to be relatively self-sufficient in four to six Earth-years. Its purpose is to provide a means to explore Mars in depth in a widening reach from the base. We would also use it to test how well living things might adapt to Mars within the life-support system of the base.

"You notice that I refer to this as a conceptual plan. Our first pass is to develop what amounts to a proposal we can use to gain NASA approvals and financial support. I am now going to work through general areas of responsibility.

"Tim will use the *Ares I* launch vehicle to get the crew module to Mars. He will be responsible to develop the concept for the crew module that will deliver the robotics, equipment, and supplies to the site for the base. He will also develop the requirements for the base itself. Consider this something like transporting the materials to construct the International Space Station and then supplying it—except this base is very much farther from Earth."

Tim shook his head and said, "We have a major problem with our heavy-landing craft. On Earth, we burn off energy with ablative tiles in the thick atmosphere. With Mars's thin atmosphere, we cannot do that. It must be done with gradual high-altitude aerobraking, substantial descent rocket burns, then some novel design drag parachutes, perhaps with two versions in sequence. This is going to take some serious study."

Louise said, "Tim, you will also be limited to a CLV compatible with the *Ares I* limitations. These are just transports, but you'll need to incorporate life support for plants and animals in the CLV. The original concept was for the CLV to carry four people. Since we will need substantial equipment and supplies on Mars, we suggest a crew of only two.

"Joan, you will be responsible for finding a site on Mars with water and other critical minerals needed to make the base self-sufficient. You will also be responsible later on to mine and refine these minerals. We suggest you explore for water and other minerals in the region just south of the fossae on the northeastern edge of the Tharsis Bulge as you see on your map. You must map the site in detail and overlay that with all of your mineral exploration findings. But originally, you must work with Rob on the initial outside setup."

Joan said, "The Mars Orbital Reconnaissance mission using ultrasensitive optics has given us some new information. We have some promising sites in prospect."

Louise said, "Rob will be responsible for describing how the construction of the base may be achieved using almost entirely native resources. Since the base will eventually grow into a sizable facility, this is a big deal."

Rob was having his first experience with underground and AI equipment. "I have another concern from the outset. We need to grow this base from scratch. We will originally need to produce all of our environments on the surface before we get underground. This is a two-stage project. We also must have substantial power for that purpose. Joan, have you something in mind for that?"

Louise replied, "Yes. In fact, you and Joan need to team up to present that concept. She will be responsible for obtaining water and using water to produce the atmospherics and fuel you will use in the underground base.

"Mary will be responsible for the living things. We want you to apply your Ames experiences to the real place with development of a food chain. We've a few ideas you should consider incorporating into your work. Consider the expertise at Ames as at your disposal. We suggest you begin with hydroponics once Rob gives you an underground habitat space. You will need to add rapidly growing algae, rabbits, chickens, and tilapia early on to get a feel for their adaptability. You should also work on developing viable soils by composting with bacteria and the likes of termites and then mixing in animal waste.

"Joan, we suggest you explore for water and other minerals in the region

7

just south of a cluster of fossae you see on your map. You must map the site in detail and overlay that with all of your mineral exploration findings.

"Rob, we suggest digging the base into the rim of an impact crater. That should be located in the region of the fossae and near the water supply. We suggest lining the rooms and passageways using stones molded into shaped blocks. They would be fused into solid stones using the thermite process. Here are some samples we've molded. You'll need life support for plants and animals within the base.

"You must also assemble many processes—first working with Joan to test for water and then drilling a well that will need to melt the ice before bringing it to the surface. Since the water will contain all sorts of minerals, you'll need to distill the water, but keep the residue for its soluble mineral content and to search for any evidence of early Mars life.

"You'll initially use solar panels for power. You'll use that power to extract oxygen and hydrogen from the water and freeze carbon dioxide out of the atmosphere. You can then produce methane fuel for internal combustion engines.

"Rob, you'll immediately assemble a nuclear power plant since our solar power will be very limited and will not work at night. The power plant is in three parts and will be transported in components. Air-transported nuclear power plants were used a half century ago by the military for arctic installations. Start with their ideas."

Louise then opened the discussion for questions.

Rob asked, "Why now? What is our motivation?"

"We've a concept that's radical and has not been considered ... that's using AI and small craft and leaving out any return to Earth vehicles. If we can economically build a base on Mars, then it opens the door for other possibilities. Now ... because Earth and Mars oppositions are decreasing, the distance traveled between the planets for the next number of years will be least distant when we should be sending people. We're approaching the least travel time we will have for some time to come. That's the least zero G travel time for people and animals.

"There's one other consideration. Industrial processes and vehicles on Earth are producing more carbon dioxide and greenhouse effects than ever before. No one can control it. The consensus has developed that global climate change is upon us.

"One thing is certain. The only industrial fuel source that's relatively abundant and avoids atmospheric pollution and global warming entirely is nuclear. If we set the example on Mars, maybe people will take a hint. Anyway, we want to have a toehold on another habitat for humanity just in case things get nasty."

Tim caught the implication. "Will we eventually be sending people? I need that for design of the transports."

Louise said, "Design the life-support space of the transport large enough for sheep and goats or the equivalent of two people. As with the Lunar Excursion Module, everything will be highly automated with artificial intelligence robots, AI-operated vehicles, and AI-controlled construction equipment.

"Remember that this is a very long voyage in zero gravity. We will be transporting either eggs near freezing or fully grown animals. Adult animals must be given a centrifugally produced artificial gravity. Eggs will be hatched after reaching Mars. Young animals will not develop normally in zero gravity.

"We do not know where this might go, so do not box yourself in. Certainly you must not include discussion of any prospect of human transport in your design documents or verbal discussion with others outside of your group."

Gus stepped in and said, "That's very important. Treat any prospect of human transport as extremely confidential among yourselves. Before that could be considered, we must prove the viability of animals and plants living inside of the base as a controlled artificial biological living space. Also, the base should be built into a crater wall similar in size to Meteor Crater in Arizona. That's two miles in circumference so Rob will have plenty of room for interior construction. You must build spaces for industrial processes and the living things. Of course, there is one more reason for going underground. The sun occasionally experiences intense solar storms that can be fatal to any exposed living things so the crater's wall must provide a shield against the solar wind, cyclic peaks, or otherwise."

Gus and Louise left the room. They had deliberately shocked their prospects. In effect, they wanted to force an argumentative situation. The question was whether this mission was plausible or not. Tim and the others were left with the prospect of a huge and unreasonable project being dropped in their laps.

Tim said, "I just don't like this. They gather us here to consider an alternative and extreme approach. They have apparently found their current plan to be impractical, probably because it is so expensive. I expect they want to see if we can find a way to go to Mars on the cheap."

The others were also perplexed. Mary seemed the most concerned.

"I've been working on ways to sustain life on Mars in very confined spaces. There are theoretical ways for doing that. I've done it in my lab. But they want us to produce self-supporting life support for an indefinitely long period of time. That time frame concerns me. There must be a highly redundant and safeguarded facility because equipment fails. Guys, this is bullshit. What are these suits thinking, anyway? They're playing us for suckers, putting our asses on the line just to see if we can make things happen."

Joan nodded. "On the one side, Gus is clearly sticking his own neck out. He must really want this to happen. And he must see the prospects for a well-financed project as pretty dim. On the other side, we're being asked to at least put our professional careers on the line and who knows, we are young and athletic so we could end up being the ones they want to send on this crazy trip. This is right out of science fiction. We're being asked to do the impossible on the very long shot we might succeed."

Rob had been listening intently and said, "Tim and I are trained to accept high levels of danger if success has a major payoff. I want to see how they approach the project when push comes to shove. But you need to see this as a trap. As you are drawn in deeper, you will become entranced with becoming the focus of the ultimate human adventure. Gus is downright enthusiastic, but Louise is much more cautious."

Tim half smiled. He said, "I just want us all to agree that we will be very open with each other. We have to look out for us. We also need to share our own projects because, even if we are not professionally proficient in the others' activities, we need to express what we are doing verbally just to go through the mental exercise. From here on out, we must really depend on each other."

With that, the team initiated a custom that they would follow throughout their association. There was work and there was play; all work and no play would not only become dull, it would destroy relationships. They needed to stay focused on the job and have creative bursts of discovery.

*   *   *

The four of them headed for greener pastures and began mingling, making themselves at home at the local watering hole.

They settled around a table and ordered a round of drinks. Joan was ready to explore her companions. She said, "Let's each give something of our lives. We are not just what Gus has told us. Yes, I'm a geologist, but my interests run broader than just being a rock hound."

They all laughed.

Tim said, "I flew jets for the air force and took a few loops around the Earth with the shuttle and the ISS. I was once married, but that didn't work out. No children. My parents and two sisters live in Florida. I like to hike, particularly in the national parks. I've explored most of the Western parks and find them spectacular. My Irish Setter usually goes with me. I'm not in a committed relationship." He smiled at the two women and they smiled back.

Mary said, "I also get out of my greenhouse. I'm heterosexual, but I've never been married. Since I've been working in California, I like to spend time at the beach, but I'm not adverse to the likes of Yosemite Park. I also like horseback riding. Do any of you know of a stable?"

Joan was not to be left out. "Mary, you have company. I'm originally from New England and I've hiked the entire Appalachian Trail. I've had two serious boyfriends, but nothing just now. I also enjoy the natural world, particularly the mountains."

Rob had a devilish smile. "It seems Gus researched us a bit personally. We are all outdoorsy people that enjoy the opposite sex. I like to build things and have never been married, but that is more a matter of circumstance than by intent. Perhaps we should head to the beach this weekend."

Everyone jumped at the opportunity. Tim offered to provide group transportation.

*   *   *

Gus and Louise joined them the next morning around the conference table. They immediately sensed a change in the group. Gus looked at Tim.

"Do you have any questions so far?"

"We have many questions, but first we want to continue with your conception of the project. You seem to be going in a radically different direction from what NASA has been considering. After you've completed your discussion, we want to hear you explain why you're doing this."

Louise nodded and said, "Rob, you'll eventually need to incorporate outlying operations like mines, refining, and a landing and spaceport location for over a dozen transports. While these have no immediate prospect of going back to space, they will be valuable assets we must preserve."

Tim was testing the waters. "Will I need to design a way to bring things back to Earth?"

Louise said, "Not at the outset."

Mary asked, "How do you plan to manage all of this activity of construction and later the operations at the base?"

"We have Global Positioning Satellites orbiting Mars and a Mars Comsat linking to ground operations on Earth. These satellites are in geostationary orbits. Tim would position Omni locators around the work area and all physical objects would have identifiers so these can be found quickly and readily by AI robots and AI-directed equipment. The equipment will have lasers to guide their activities using position markers.

"Beyond that, we anticipate a JPL-type control room that would operate twenty-four-seven, staffed by experts like you. These controllers would direct activities and monitor their progress. JPL would continue to manage the space flights, including approach, descent, and landings as well as working with Joan in exploration. Some of their people would be expected to staff the new control center after helping Tim design and construct the interplanetary transports. We expect the AI-controlled activities would be directed using macros issued by the controllers."

Mary's concern showed. She said, "You know that working with living things in an enclosed space can be very sensitive. The original Geosphere in Arizona was extremely difficult to control. There was leakage of the atmosphere and there were gases issued by the living things. Temperature is not easy to manage here on Earth. Underground on frigid Mars sounds more difficult. Hydroponics require grow lights, water flowing uniformly in troughs, and maintaining levels of nutritious chemicals in the troughs. You must keep power to the grow lights and the lights themselves working all the time, which sounds like a nightmare without people there to keep an eye on things.

"You need to harvest the plants and seed new plants. You need to feed the animals—although it sounds like you want growing plants to feed the rabbits, chickens, and fish. All of these animals are prolific; they reproduce very rapidly. What do you plan to do with the offspring?"

Rob asked, "Mary, can you really grow food in such a tight space?"

"Sure, down to about three cubic meters can grow food to feed one person. But if you can give us decent-sized interior spaces, I could do a lot more. If I were to be playing *Swiss Family Robinson*, I would want to send along some animals on the first robotic ship to see how well they fare in the environment we build. Please tell me you can build us some generous living and workspaces or the diet could get very dull. It would be really great it you could have that tested and waiting when the first people arrive."

Gus said, "Mary, that's why we need you. You're the best. All of you are the best at your professions. We want you to beat this to death, both pro and con. Down the road, we'll need to make some decisions. Do we leave out the living things? Do we want more or less animals? Which animals should we include? How much space and what equipment do we need? Do we need to send people to control the operations in person and in real time? Is the prospect of sending people out of the question? We need to answer a lot of questions. That's why we can't answer your questions. You must do that for us.

"Most important, we must design our effort in attainable pieces. Whatever we do, we must maintain a promising success level or we just will not be given the approvals or the funds for the project.

"You're the ones who will determine if mankind is ready to go beyond our home planet. We're at the beginning of what must appear to be a succession of increasingly more ambitious undertakings built one upon the other. The first two steps must be to develop this conceptual plan and to redirect one of the next two missions to Mars for 2012 to the fossae region, looking for water. As far as anyone else should know, you're doing long-range planning for Mars exploration. Sending transport ships and beginning AI-controlled construction might begin as early as the following Earth-Mars opposition, two Earth years later. That's all."

Chapter 3:

Comprehension

Gus and Louise left them alone in the room.

Tim finally said, "We were right. We've been handed the biggest exploration event of all time."

Rob had been playing with the stones. "They've been working on this already. In fact, Mary has been one of their unwitting players. Furthermore, someone has done some creative work with these stones."

Joan was toying with her sketchpad. "You shouldn't feel surprised. You're just the focus. You need not do everything. You will have all the resources of the organization at your disposal. If you need something done, you should identify it as a task and then get approval for someone else—in NASA or the educational or business world—to do it for you. This is not all on your shoulders. JPL or Ames or USGS or most anyone else may become your partner. That's particularly true for you, Tim."

Mary said, "My question did give us some focus. One thing I got in their response was that this is definitely a humans go to Mars project, but nothing like anyone has suggested until now. Joan, I'm still overwhelmed by the prospect."

The concept team shifted into brainstorming.

Rob said, "What I don't know is how much of this can be done with robotics from here. The critical part is getting the electrical plant operating. Nothing substantial can happen until we've got power. Along the way, we demonstrate that we can produce enough water from wells and distill it to make it pure."

Tim felt enthusiasm despite his reservations. He said, "Everything else could wait until the next planetary opposition when we send another electrical plant as backup and perhaps four colonists. To really get things jumping, we'll need the people there to expedite matters."

Joan was resisting the plunge into fantasyland. She said, "Careful about the people. I agree, but that's not ours to insert into a plan at this stage."

Tim said, "We could use a modified *Ares* rocket that has a Crew Exploration Vehicle. I suggest we rename our version the Base Development Vehicle. The BDV must be capable of independent landing on Mars—with a relatively heavy load, which is the real nut we need to crack. Rob Manning at JPL has managed the most recent landings on Mars and foresees some very serious problems. Mars has more gravitational attraction than the Moon and therefore is going to suck the lander down rather precipitously. The atmosphere is enough to cause a problem since it's not dense enough to provide ablative energy conversion to heat. The parachutes used on light landing modules help, but they cannot slow these heavier loads enough for a safe landing.

"So we need to find some form of drag to slow our BDV from as much as five thousand miles per hour to something we can handle with steerable rockets—say a reduced speed of one thousand miles per hour. One proven approach is to use aero-braking, which inserts the BDV into an elliptical orbit around Mars that skims through the upper atmosphere at the low end.

"Another concept uses an array of sturdy, doughnut-shaped, drag-inducing objects trailed behind the lander that are released for the precision landing. We need to brainstorm this idea, test it in a wind tunnel, and finally test the real thing behind high-altitude rockets where the Earth's atmosphere is relatively thin.

"There'll be no other vehicular components for the ship. We'll be doing essentially what they've been doing all along with the explorations—except we design for this exotic rocket landing."

Joan jumped into the conversation. She said, "We need to send at least one standard research craft to find the water and at least four fleets of four BDV—one mission for the robotic base setup and three more missions with people, equipment, and supplies. If that first robotic mission is successful, it will result in all the basic operating facilities and some plants and animals living in the new base. The empty base will provide living space for the first people and more varieties of animals and plants from the subsequent fleets. With that in place, the project should sell itself."

Tim leaned forward over the conference table, got up, and began pacing. His voice rose as he said, "I want to keep the transportation in focus. Constructing the additional ships will be much less costly than the first one. Also remember it was *Apollo 11* that first landed men on the Moon, not *Apollo 1*, and there were a total of eighteen Apollo ships that were built. Each of our ships would have two modules—one to power the ship to Mars orbit and the

BDV payload module to land on Mars. The learning curve should be familiar. Unlike Apollo, we do not need to launch our ship back into space and we do not need a separate command and control module."

Joan said, "You're still sliding past the critical element. We need to find all those indigenous materials on Mars fairly close to each other. Abundant water is essential. We must make a list of important indigenous materials and then find them—at least the most critical ones."

"Impure water will do us very little good. It must be distilled to get the pure water that is needed for much more than drinking or plants. It's needed for steam generation in the power plant.

"Pure water can also be split into hydrogen and oxygen which is our best initial source of hydrogen. We need hydrogen to form fuel for internal combustion engines. These gases are very convenient to have around. Commercial production of hydrogen on Earth begins with hydrocarbons which we will not find on Mars. We can eventually get oxygen through photosynthesis of carbon dioxide by vegetation.

"Other natural resources are needed. The Mars atmosphere can be compressed and cooled to produce dry ice, which separates out the large proportion of carbon dioxide. It already happens naturally at the poles of Mars each winter.

"Removing the carbon dioxide from the native atmosphere leaves a mixture of mainly gaseous oxygen and nitrogen. That mixture can then be further cooled to liquefy the nitrogen for separation from the small amount of other atmospheric gases. Nitrogen is needed to dilute the oxygen in the habitat atmosphere. The cold natural atmosphere can be used to advantage in these refrigeration processes."

Mary asked, "Why are we producing methane? The answer is that burning methane and oxygen as fuel produces carbon dioxide and water, which are non-toxic, so these internal combustion engines can even be used around the habitat while it contains just Mars atmospherics. It could not be used in a balanced habitat atmosphere."

Rob directed his next question to Tim. "How do we conduct this search for water and how soon can we do that?"

"I would initially use the same delivery platform we're using for our current exploratory missions, but I would reconfigure their cargo to support water drilling and to carry our new experimental miniature aerial platform. That's already planned to conduct aerial photography leading to the detailed, large-scale mapping we will need for development of the base and the nearby region. We also would employ the latest robotics.

"As to when, I hope we could hijack one of the two missions in 2012.

The water drilling process uses the same equipment we would use to retrieve underground soil samples. Nothing exciting is needed there."

Rob said "Budget-wise, this could be downright frugal. The hard part will be selling the concept of leaving a crew on Mars. We wait for events to push us into doing that. We should be able to do this remote exploration in 2012. Even if we need to push back the first *Viking 2* launch date, we would still be doing quite well. Remember that the Apollo project to send men to the Moon began in 1961 and was accomplished in just eight years."

CHAPTER 4:

FORWARD IN STEP

TIM DISCUSSED HIS professional thoughts with their team during the next morning's planning session.

"We're approaching this project from both ends and will work toward the middle. Joan and I will work on the exploratory mission with the JPL people. That will be out in the open. Rob and Mary will be working on the requirements for the design of the base. We are over three years from our Ares mission to actually begin base development, but we will need all of that time. The base planning is to be confidential."

Rob said, "Our immediate task for the unmanned base development mission is to produce fuel for our engine-powered equipment. We then must use that equipment to assemble a nuclear power plant with capacity enough to power small-scale industrial processes. Then, given relatively abundant electrical power and related heat production, we use that to power our various operations and to drive chemical processes."

*   *   *

The only distraction from their work was themselves. It had been evident from their first meeting. They were all singles who normally led active social lives. Now they were absorbed in a secretive project with no outside distractions. It was only a matter of time until they could no longer ignore the obvious.

Joan and Mary were enjoying a break together.

"You know these two men we're working with aren't half bad," said Joan.

"Too bad we're working together," said Mary.

"I've been wondering if that applies here."

"There is another possibility. They did put us together, and I don't think they accidentally had two women, two men—and all single. Do you think they're waiting to see if their matchmaking will come alive?"

"You've got a point.

Mary's smile was a bit devilish. "Are you timid?"

Joan's eyes sparkled. "Not unless you are."

The men walked into the room and Mary and Joan whispered simultaneously. Rob and Jim could tell from their body language that something had changed.

Rob lowered his head a bit and said, "You look like you're reaching into the cookie jar."

Joan made a come-hither movement. "In this case, it takes four to tango."

Tim took a half-step backward. "Forbidden fruit?"

Mary held her hands open. "Perhaps it's not forbidden. After all, they arranged this ... you know ... two and two. Maybe they want this to happen."

Rob took the bait. "We're remarkably compatible. Please correct me if I'm misreading our current relationship, but we do seem to get along quite well personally."

From that point, the four began to take a more outward personal interest in each other. The working relationships had them paired and the pairings were natural. Tim and Joan were more outgoing and athletic. Rob and Mary were more focused. They went out to dinner as a group and then tended to find opportunities to talk as couples about their lives and interests. Intimacy would not be denied.

*   *   *

Their work progressed as time passed. Joan was responsible for the minerals and their processing.

"Here on Earth, we would naturally search for aluminum ore, bauxite, and use lime to produce alumina or aluminum oxide. Here we find bauxite in commercially rich deposits. Perhaps we will find it conveniently on Mars, but perhaps not. *Pathfinder* examined the Chryse Planetia surface of Mars back in 1976. That surface material was not rich in aluminum by any means, but it was almost 10 percent aluminum oxide and over 11 percent iron oxide.

If we find rich bauxite conveniently, then we go that direction. Otherwise, we use the richest region of aluminum oxide in the surface dust and go from there. The point is we want to get things moving as soon as possible, and we don't waste a lot of time looking for the best solution when we have a viable one looking us in the face.

"We will begin with a single small refinery bath and a single smelter pot. In time, we will increase the size and number of these, limited primarily by the electrical output of our small nuclear power plant.

"The next step is to mix powdered aluminum with iron oxide and other materials to obtain Marscrete. The aluminum can also be formed into all manner of shapes and useful objects."

Tim tied it up. "All of this must be accomplished before we send manned ships to Mars. We must have a life-supporting space to make a habitat possible." While their work schedule was intense, the social time was fully intermixed. What had begun as a dating arrangement soon became a full social calendar. They frequently went to the Texas gulf coast for beach time and on other occasions explored mountainous terrain in the national parks system. They were all athletic and were soon hiking wilderness trails as far away as the Grand Canyon and the Yellowstone – Grand Teton ecosystem. That was Tim's interest, so he became their guide. This activity guaranteed they were in excellent physical condition.

Their exercise stimulated other interests that were readily accommodated. Rustic cabins and pop up tents soon became intimate enclosures on virtually every opportunity. They had naturally paired according to their interests and without any outward understanding they were strictly monogamous. They knew their work life could not stand the strain of a more complicated relationship and they clearly were not about to behave in such a way as to risk their remaining on the Mars project and possibly its growth into their becoming pioneers.

Their project became more focused as the launch of the 2012 mission approached. Tim discovered they needed to minimally modify the payload intended for the mission. Joan identified and redesigned the remotely controlled drilling equipment for subsequent operations once water was acquired. They worked openly with JPL.

Rob's design for the base and Mary's life sustaining facilities were not needed in the exploration phase. They were working with a team of six people scoping out the requirements for a Mars base under various assumptions.

Tim and Joan faced a frustrated JPL team that was none too happy to switch mission goals and target locations. A lot of work had gone into their project only to have it hijacked for another purpose. Still they all were Mars addicted through and through. No one lacked enthusiasm, particularly since they were smart people and they, too, were speculating on where this new direction may take them.

The days seemed to fly by, but the nights were another matter. They moved in together, but kept their original places for appearances. The time together alone was just as compatible as they had hoped. Their casual interests matched as well as their professional and their sexual exploits only matured. Mary talked her friends into visiting Yosemite National Park and to practice some rock climbing. They ran into some serious climbers and listened to their tales of hanging from the sky. El Capitan was as spectacular as could be and Half Dome was equally impressive. A time would come when they would reminisce of such blue skies and green landscapes.

Joan's redesign of the survey mission went very smoothly although fitting in all of the equipment for the water and chemical processing required leaving out some pet experiments that sorely disturbed their designers.

Test upon test were conducted in simulated environments in Alaskan research areas. The drilling was specifically designed to reach known aquifers and actually retrieve subterranean water and then process versions of the adulterated water using the miniature testing and distillation equipment. This equipment would all be bounced down on Mars so there were interminable impact tests from high towers and drops from aircraft. That was followed by realistic remotely guided ground operations with the communication delays encountered at interplanetary distances. Sending messages to Mars required a ten to twenty minute delay before a radio response. This was not new for the JPL team.

Tim and Joan had turned over the operational control of the mission to the JPL team once they were assured it should all work okay. Joan continued to monitor progress.

NASA had been working on an *Ares* family of rockets to carry men back to the Moon and ultimately on to Mars. The first stage had been launched for a two-minute flight from Cape Canaveral October 28, 2009.

What Tim needed was the *Ares 1* version that would carry cargo and he needed to move it up to carry the first cargo to Mars on a very tight schedule. In effect, he needed to use the first *Ares* to carry the equipment and AI robots that was needed to explore the prospective colonial region for water and mineral resources.

Tim had gathered his own technical team to begin confidential design of the modifications that would be needed to carry the specific cargo load for their conceptual mission. That was when he realized the technical enormity of his project.

Mars did have lower gravity than Earth, but much more than the Moon.

Inertial mass was what had to be slowed to a stop as the ship descended and that mass, except for fuel being burned, remained constant. Their method was approach into orbit, separation from expended propulsion rockets then descent using a trailing braking mechanism and landing using only parachutes and landing rockets.

Louise held regular weekly reviews with her development team and all monthly meetings were joined by Gus and quarterly by Dave for all-day thorough reports.

Dave joined them at a late 2011 meeting. Louise moderated. But first she had some personal comments.

"We are very happy with your progress, and I do not mean just professionally. Unless I misread entirely you have found compatibility on all levels.

"In project summary, we are making unexpectedly good progress and are actually about two months ahead of schedule. The equipment for the search mission is not only constructed, it is fully tested for function including the Alaskan drilling tests. JPL has very positive results on the approach-descent-airbag landing tests as well. We're moving to the impact landing tests now to be sure our new equipment can withstand the landing stresses. JPL has thoroughly surveyed the first landing site which is a bit tricky. It lies alongside a monster bluff in the Kassei Valles. The fossae landing site is more open, except we do not want to end up in one of the depressions.

"Tim freed up from the survey project about two months ago and is now working with his new team on the modified design to see what can be salvaged from that experience. He will fill you in. Joan has some information on the geological conditions at the landing sites and is encouraging us regarding expected minerals at the fossae site. She has also been working with Rob on expected construction conditions. Rob found that the rubble pile at the site of the bluff at the valley site is a complication."

Rob spoke first. "You speak of Mars as a distant world with features described in astronomical terms. I will be speaking of Mars firsthand as an engineer, much as if we were conducting our work here on Earth. For example, I will refer to the Tharsis Bulge as the Tharsis Highland. I will refer to geologically ancient stream beds as though they were just dried streams in a modern sense.

"We'll be making certain environmental assumptions. For example, we will ignore the effect of the very thin atmosphere on construction, which is to say the wind will be powder puff light. We will, however, remember that all living things must be encapsulated in pressurized suits or containers for their outside movement.

We'll assume an 80 percent Earth atmospheric pressure as our inside pressure for living and working space, with a 25-73-2 oxygen-to-nitrogen-to-carbon dioxide mixture. That means breathing will inhale the same amount of oxygen and plants will experience the same availability of carbon dioxide for photosynthesis as on Earth. This nominally lower pressure will allow people to suit up without excessive delay for outside activity. Unlike space projects, we do have a need here to maintain relatively high air pressure or else our breathing subjects will not be sustained and growing subjects will not develop relatively normal lung capacity.

"Outside, the thin atmosphere provides little convection resulting in only a couple of feet of atmospheric solar warmth adjacent to the surface. We should also appreciate that overnight temperatures even in their summer at our latitudes are downright arctic. We will consider the effect of extended dust storms obscuring visibility. These storms would also greatly reduce the electrical generation by solar panels and could make cross country transport very dangerous.

"We will not need to allow for drainage since there is no rain nor will there be any roofing as such, except as enclosure to protect from dust and to contain atmosphere. We do need to eventually provide raised and paved running surfaces in order to avoid surface dust damage to vehicles and as a tarmac for landed spacecraft.

"Since both prospective development areas evidence erosion we anticipate that we can move soil readily by excavation, including digging into the crater wall.

"Are there any questions so far?"

Gus responded. "How are you managing the extreme cold?"

Rob continued. "Most construction will be well underground. All heat originates from the nuclear reactor but the distribution of heat will be from the power loop using the spent steam after the steam turbine. That will go to the living and processing facilities space. We can also capture some of the heat from any thermite processing. Beyond that we will use electrical heating. Further along, we could use some form of conventional glass fiber insulation on walls.

"That brings up the matter of Marscrete testing. We have used your work as a beginning. We have formed many block shapes and we have erected many structures both outside and as linings to inside spaces. We have particularly formed domed shapes underground and subjected them to destructive testing. We have also used a Marscrete paste to fill the small gaps between blocks and have fired them to create essentially continuous walls. We have tested the walls for pressure tightness, essentially to see what it takes to make them thoroughly sealed. We have tested the optimum thickness for walls and domed ceilings."

Gus returned to his instructions. He said, "Okay, there is one other condition that is somewhat overriding. We must do everything possible to avoid wear and tear on garments, spacesuits, equipment of all kinds, electronics, and particularly anything that moves. Surface dust is abrasive and can be very destructive. Everything taken outside must be cleaned upon taking it inside. That includes all equipment and clothing. You will be able to manufacture cloth objects, but other objects will wear out, so you must be very careful with objects that cannot be replaced on Mars.

"We'll need to convert minerals into useful materials and grow a wide variety of foods and vegetative materials. So, before I go further, Mary will discuss our needs to support people, plants, and animals."

Mary smiled to offset Rob's serious demeanor.

"You're already very familiar with atmospheric needs, water, and our general thoughts on hydroponics. We will need sensors and automatic controls aplenty. We're also planning on animals. They will prove the viability of animals living on Mars. We plan on a fairly wide diversity of species, focusing on animals that are small and highly reproductive.

Mary continued these thoughts. "All of these activities will ultimately be connected by light and relatively narrow Marscrete roadways and walkways—except where heavy transport may be anticipated. People could move about outside in pressure suits, using personal, electronically stabilized, two-wheel Segway PT" Personal Transporters. They allow ready boarding by the operator. Walking for appreciable distances in bulky pressure suits can be very tiring—and can wear on the suit.

"Inside, we carry ballast distributed about our bodies to maintain something like normal body weight to prevent atrophy of our muscles. That will not help our hearts, so we must perform considerable exercise that stimulates blood flow.

"Joan and I have worked up a list of essentials. It's longer than I would have hoped. We are very aware that there is no hardware store nearby where we can pick up items we may have overlooked or which may break. The two years until the next delivery can be a very long time."

CHAPTER 5:

VIABLE PLACE

ROCKET LAUNCHES WERE old hat to Tim and the JPL team. For Joan, Rob, and Mary, it was another matter. All four flew down to see their rockets launch, carrying their exploration modules. Even Tim was into the action, giving a launch sequence narrative and a tour of launch control at Cape Kennedy. As the launch count reached the final countdown, Tim wrapped his arm around Joan's shoulder. Each was imagining what it would be like to be riding atop the nose cone. Mary was familiar with NASA activities, but she placed her hand into Rob's tense grip. Could their dream ever come true?

Mary whispered, "Who are we to be involved in this outreach across the heavens? I'm glad we found each other. I trust you."

Rob turned and faced her. "And I trust you. I can't imagine being without you."

Tim broke the reverie. "I'm glad we appropriately named our new mission rockets. Lewis and Clark were seeking a passage across a continent into unexplored territory for humanity."

The first exploratory rocket, *Lewis*, disappeared into the distance. Its payload would bounce down after its six-month interplanetary journey into the northern Kassei Valles near its natural estuary into the eastern sea. There the valley was broad, water eroded, and with high steep walls.

The second rocket, *Clark*, was staged to launch two weeks later, but it was targeting a plain adjacent to three huge sinks called fossae. They were located along ancient geological fault lines. Presumably hot steam escaped from the fault lines like Yellowstone fumaroles, but in this case, the water was not being replenished. That eventually caused the overlying ground to collapse into the vacated space. They were betting that there was still water underground off to the side of the fossae.

The sites would be explored using an aerial vehicle the team called a humbee because of its small size and its maneuverable flight characteristics. Hummingbirds and honeybees could hover and move laterally in all directions.

The humbee was made with an exceedingly light helicopter rotor frame employing two coaxial synchronized counter-rotating rotors of four blades—each propelled by rotor-tip rockets. That design was necessary to neutralize the torque produced by a single rotor and to generate maximum aerial lift.[1]

The rotor blades looked like razor-thin dragonfly wings. Methane and oxygen fuel were driven by centrifugal force down the rotors' leading edges to needle-like micro rockets. The tips of the rotor blades were connected with each other by a monofilament to stabilize the tip-path-plane. From the ground, the ship looked like a small flying saucer.

The Lewis was landed eastward down the length of the valley and made a remarkable approach and bounced down, enveloped in a cluster of durable balloons. It bounced very high, maintaining its landing forward momentum. Its second ground contact was off to the side, striking some boulders that caused a less spectacular ricochet. That erratic behavior was repeated for six more bounces before it finally stopped.

None of this was viewed at Earth Control. The balloons deflated and left the equipment package upright and open to the sky. It conducted a self-test of the equipment, video-scanned the neighborhood, and finally activated the GPS system to determine its precise location.

The humbee was started while the vehicle was still held firmly in place on its magnetic base. Its first local flight was followed by a search for minerals and drilling for water in the immediate area.

The Kassei site was preferred because its climate was warmer and the high walls would shield their activity. It turned out to be a disappointment. A number of useful minerals were discovered, but drilling did not find any underground water. However, the images sent back from the humbee were rugged and spectacular.

The fossae site was considered more likely to have water. The terrain first examined was the plateau just south of the three huge fossae where they hoped

---

1   *Reference is made to video of the new Sikorsky X2 with dual counter rotating rotor blades in actual flight test. http://www.sikorsky.com/Innovation. Your author's concept was developed independently.*

to find a deep underground ice aquifer. On Earth, Tim called for Rob and Mary to join Joan watching the survey develop.

The flight was Mars AI controlled. The humbee fired its rockets and the miniature ship lifted slowly from the base. The team stood transfixed, watching the incoming images on a monitor. The small intermittent views sent from the micro video camera grew from a narrow field of view to panoramic width as the humbee rose higher. The full video was being stored on the landing platform on Mars. High-resolution vertical photographs from the humbee would be the source of detailed contour maps of the entire prospective colonial site.

The humbee sent its altitude and movement to the Mars AI controller that directed it along a planned route. The natural colored, sharp photographs gave the impression of flying across the Mars landscape. The local time was midday and the ship was flying west. The photos appeared much like sepia photographs, even to the horizon and the light in the sky. The higher terrain met the sky with a sharp demarcation.

The entire team was transfixed, knowing that this was the possible payoff in the search for water.

The surface was littered with boulders; most appeared anchored in the ground and had the general appearance of a dusting of pale pink snow. The occasional planar surfaces of the boulders were edged, pronounced but not sharp. The larger boulders were irregular in shape; a few had relatively flat tops. The tops of the boulders looked light in the prevalent Mars desert; the sunlight struck the surface at about a forty-five-degree angle. The craggy sides of the boulders were irregular, but not worn from erosion. The surfaces of the ground between the boulders were strewn with stones and uneven. The boulder surfaces in the shade were distinctly darker than the side in the sunlight. The horizon outlined nearby rounded hills with occasional large impact craters standing higher than the rising surface.

The four prospective astro-colonists were absorbing the realization that this could become their new home.

Tim said, "I would have liked a better landing site."

Joan was also in a bit of shock. "Can you actually land in the middle of all those boulders?"

"There appears to be more open terrain in spots."

Rob said, "We're looking only toward the higher terrain. We don't even know if there's water here yet."

The humbee turned south into the sun. The glare was brilliant despite its apparent smaller size when viewed from Earth.

Mary noticed lower terrain on the horizon with fewer large boulders. "Hang on. We'll be facing the old sea as we turn east."

When the humbee turned east, there were fewer boulders. The distant horizon overlooked the nearly flat Acidalia Planetia, the ancient northern Mars seabed.

Tim almost shouted, "This is more like what we need. We can land here."

Rob was visibly relieved. He said, "This is manageable. We can clear a roadway right away and use explosives to make the road fairly straight. Do you see the crater? It's just what the doctor ordered."

The photo mission turned north about a kilometer from the one-thousand-meter-wide impact crater. That was already designated Genesis, their apparent home to be. The ship crossed the lip of the nearby southern fossa. The ship continued on overflying the fossa and then turned east. The depression bottom dropped abruptly over three hundred meters. The craft continued eastward, maintaining altitude. The sides of the fossa showed rubble piles—even to the eastern end. Such rubble from collapsing mesa walls was very common in places such as Utah's Monument Valley.

Joan was already very cheerful. She said, "So far, so good. Now we need to find water and those critical minerals."

Everyone was avoiding the obvious; no one wanted to break the string of good luck.

The humbee returned to its base platform, landed with precision, and was anchored. The small fuel tank was recharged before the next trip. Then the ship began a detailed survey, flying low and using a miniature Thermal Emission Spectrometer. The TES could sense the mineral composition of the surface using the infrared emission of the minerals. The humbee would hover over interesting locations, dusting off the surface to better expose the surface before taking measurements.

The data already captured in the AI processor in the base was used with overlapping photographs to contour map the surveyed area in considerable detail. Next, a small rover was moved to strategic locations. Survey benchmarks were sunk into the ground and a few Omni directional radio homing beacons were anchored for vehicular navigation in the sea of boulders. These were to be used for navigation of mobile equipment. The mineral findings and prominent landmarks were located on the map.

The humbee also examined debris piles at the base of the high cliffs and the cliffs themselves. Together these provided the first local detailed geological survey on Mars. Then the survey information and map began transfer to Earth control. This was not a high-speed process; it took hours while the mineral findings were located on the map.

Next began the most critical operation of all. Everyone was gathered around the monitors that would follow progress in drilling for water. The first of three sites was slowly deep drilled near the prospective crater. The core samples were carefully catalogued for identification of mineral location.

The drill slowly penetrated the depths of Mars for the first time. Over and over the drill was extracted, the sample removed, and the drill sunk back into its bore hole. Everyone was monitoring progress as the drilling reached a hundred meters without any luck. The team knew this was still relatively shallow to find water. Each new sample was photographed on Mars and scrutinized intently.

Finally, at 150 meters, there was a change in color of the specimen—but nothing definite. The next sample was different. It was not clear, but testing showed it to be murky ice. The sample was placed in an oven and heated. It melted and the liquid sample was tested for content.

Tim was reluctant to declare success. The following samples continued to produce ice as the drilling penetrated deeper. Finally, Mary couldn't hold back.

"Come on, guys! This is it! We found the most exciting discovery of all—this is our Holy Grail!"

Rob reached into a nearby refrigerator and pulled out a bottle of champagne. He excitedly began unwinding the wire over the cork and began the gradual process of extracting the cork. He pushed the cork up with his thumbs so that it would fly across the room. The pop was loud and clear. Glasses were filled and the entire room began opening more bottles. It was like a Super Bowl victory celebration with champagne being poured over the heads of Gus, Mary, Joan, Rob, and Tim—and everyone else they could find.

The media would not be denied. They invaded the work area. Meanwhile, the AI equipment on Mars was methodically bringing up more and more ice samples. That continued until reaching 300 meters depth. The aquifer was a full 150 meters thick.

The drilling equipment on Mars was moved to a planned location five kilometers away. The drilling there hit ice at a slightly lower depth, but with just as much thickness. Then the drilling began at the third location a good eight kilometers away from the other two with the same results—they were on top of an underground Martian lake.

This was the first encounter with significant quantities of the ancient water on Mars. The water was tested optically and chemically for any indication of embedded life forms. Everyone was hanging on the possibility of finding any indication that Mars had once been home to living things—even of the most primitive form. The water was disappointing in that regard, but it contained dissolved minerals, particularly salt, in concentrations even stronger and more diverse than the water in Great Salt Lake.

Water samples were tested for content that could help explain the geological history of the region. Some of the contents proved potentially useful, but required considerable chemical processing. The water was heated and its vapors condensed, thus distilling the brine into pure water. That water would prove necessary when the first *Viking 2* now named *Thor* arrived two years later in 2014. All living things that might be brought to Mars would require this pure water.

The other primary use would be to produce steam from the nuclear reactor to power the steam turbine. The turbine's shaft would then power the electrical generators. The most critical elements were now in place. A relatively large and very thin bladder was used to collect the distilled water on the surface and it promptly froze solid.

The humbee again proved its worth, finding minerals that were critical to developing the base. Early on, they found adequately rich aluminum ore about twenty-five kilometers west of the crater. The mapper also descended into the nearest fossa, finding even more minerals and an outcropping from the aluminum ore deposit.

JPL had worked hard to design a flexible robot. Perhaps unexpectedly, technical design learned from art. Their robots looked remarkably like the ones in the motion picture *I, Robot* and had a great variety of practical movements.

The accompanying robot and hauler set numerous survey benchmarks, designating navigation points around the crater and outward to the ore deposit. Detailed maps were made of the entire crater region and the entire likely work area for development of the prospective base.

All together, it had been the biggest discovery day yet on Mars. The entire world's media was celebrating. The exploration team, including JPL and NASA, was instant celebrities.

CHAPTER 6:

IT'S A GO

FOR DAVE, GUS, and Louise, the discoveries were extremely exciting. It was early 2013—time to bring in others to work with Tim, Mary, Rob, and Joan to compile the larger colonial plan for submission to political leaders. The water and ore discoveries had been widely followed by the media. The question of sending people to Mars began coming from the media before Dave was ready to divulge his plans. Congressional aerospace committees were asking for information almost immediately.

Dave contacted the White House chief of staff, explaining the public and political pressure and insisted that they wanted to give their information to the president first.

The three NASA planners were called to the White House by an apprehensive president who was already suffering from a very tight budget. He had been elected in November 2012 and was more inclined to manned planetary exploration than his predecessor had been.

President Jack Sorley brought the planners into the Oval Office where they found his White House chief of staff, his science advisor, and surprisingly his secretary of state and his secretary of defense. The president wanted to set a tone.

"I have my top people here because of your surprising success and my concern about where you may be taking us. Your success is becoming a problem. People want to jump off on a Mars mission that is certain to become terribly expensive. Dave, have you considered the consequences of your discoveries?"

"We have actually anticipated this possibility for the last two years while we were developing the discovery mission. We just held back because we were far from certain about success—definitely not this level of success."

President Sorley frowned at this admission. "So give me the good news first."

"What we have is all good news—provided you're willing to accept the approach we have in mind. The cost would be rather reasonable to initiate a robotic mission to Mars that would develop a base. It would be built inside a moderately large impact crater on the east side of the *Tempe Terra*."

The president had crossed his fingers and touched them to his lips. "I see. I've heard nothing about a base. What else haven't you told me?"

Dave responded, "Well, we've already been working on a heavier lift design that uses rockets to land heavier loads than with the bouncing balloon approach. The *Ares* rocket has been reactivated and will be able to transport this new base development/exploration module to Mars orbit. The Base Development Module will be able to carry components of a nuclear power plant, some construction equipment, some chemical processing equipment, and some aluminum ore processing equipment. It will take the exploration module cargo of all four of these ships to set up the basic working site. We're currently sending two quite small ships every two years.

"Please note that this Base Development Module is transported to Mars by an already developed vehicle and the module is already designed to be flexible with what it needs to carry."

"What is your goal here?"

Dave took a deep breath. *Oh, what the hell. He's got to know about the big plan someday. It might as well be now.*

"Mr. President, you've been seeing just the superficial projects that could lead us into a huge leap forward. Let's step back and take a strategic look at where we could be going. We need to table our current discussion and make some assumptions based upon the discoveries and technological advances we have already made—those that make other achievements not just possible, but probable.

"It's very likely that we can evolve our present projects into an economical project that not only places people on Mars, but creates an operational Mars colony within a decade. We would not project this to the public—just the individual projects as they rise to the surface."

The defense secretary raised his hand. "You can't be serious!"

Dave realized that he had placed himself in a critical situation. He had to punch across the colonization idea. He had to sell his project essentially cold.

"Sir, I have never been as serious in my life." He looked at the secretary of state. "First, consider where this puts us strategically in a global sense. We now know that Mars has abundant water in frozen aquifers and a wealth of natural resources. Mars can become self-sufficient, including substantial nuclear power fueled by uranium we expect to find. Those resources can become very valuable to us. It would be much easier to ship critical resources from Mars to Earth orbit than from the surface of the Earth to other places. We are running out of fossil fuels and the use of the remaining fossil fuels is a major cause of global warming."

Gus jumped in and said, "We're already facing fierce competition for energy resources and we're experiencing a rising tide of terrorism. Mars can give us a winning hand in that game—one of the few places where we still hold a strong technological lead is in space exploration. We have everything to gain by exploiting that leadership."

The secretary of state was nodding agreement. "We are in a tight situation and we do need to do that."

The president stared at Dave. This was far too large a leap for him to absorb so quickly. He had not expected quite this extreme a turn of events, but that was why he needed his senior adisors seated around the room. He said, "How would you proceed?"

Dave was reassured. He could feel the tide turning in his way. He smiled and continued in a quiet, calm voice. "We don't want to set overly ambitious goals. As we already see, the press is quick to carry us into the unknown. We must manage expectations. This is a stepwise thing. Our plan for 2014 through 2016 is just to get the basics set up using AI robots and equipment. If we can do more, then so much the better."

"So, what is the prospect in 2016?"

"We always work with redundancy or backups before setting up anything that is dangerous. In 2016, we want to send four more base development ships. Three would expand the steam turbine and generators that were sent in 2014. The other two would flesh out the base. After that, the base is essentially self-sufficient except for needing nuclear fuel and critical replacement parts."

The secretary of defense leaned into the discussion. He said, "This sounds like an awful lot of activity in a short period of time. What's the result? What do you achieve? We're already overtaxing our budget."

"We're creating an environment that supports living things on another world. We accomplish this construction and these new operations almost entirely using native Mars resources."

The national science advisor was stunned. "That is science fiction ... it cannot be real! You cannot do this in just four more years."

"Actually, all we can do in four years is place all the equipment and supplies on the surface that we need to make this possible. There will be a growing number of artificial intelligence robots. Do not underestimate what they can do with guidance from here. Even that requires that all of our projects be successful."

"When do the first living things arrive on Mars?"

"We have already taken samples of living things on a recent mission and they have done well. We take seed animals on the next ships: tilapia fingerlings, chicken eggs for hatching, and adult rabbits will all be on the next mission."

President Sorley was becoming concerned about whether he could live with all of this space activity within his budget. "Where does all of this lead?"

"We're proving capabilities using those AI robotic workers. The potential is enormous once you have a viable habitat on Mars and enough robots."

"Do you see humans actually living on Mars in the foreseeable future?"

"That is a distinct possibility, again assuming our projects are successful. Whether to do that and when is another matter. If that is to be done economically, it will depend on whether we accept certain conditions."

"Is there prospect of getting someone to go who we would want on Mars?"

"Yes. We actually have four highly trained scientist-engineers who have been working on the project and seem very interested. We have not made them an offer."

Allen Wilson, the national science advisor, chose this point to change the subject.

"What is this habitat going to be like? You can't take a lot of heavy construction equipment."

"Our plan is to dig passageways and rooms inside the wall of the Mars crater, which we call Genesis. They would be lined with solid stone arches and domes that we create using Mars resources. Here are some samples made from Mars-type materials. There would be biosphere space in the habitat, chambers for humans, and industrial and power generation facilities that must be developed."

The president said, "You make this sound so simple. Is it actually safe for humans to go?"

"Sir, nothing is safe in space or on a nearly airless world colder than the arctic. On the other hand, the base is deep underground in the crater wall and as solid a place as we can build. The crater itself absorbs the solar radiation. Mars doesn't have earthquakes or violent weather. Everything after the second year that is critical will be redundant with two or more of each type of equipment."

Gus had been silently leaning back in his chair and carefully assessing the president's facial expression as Dave laid out the plans. He decided that it was time to speak up. "That's because we've given you only an outline, a concept. As it stands now, without any modification, our plan is subject to failure."

President Sorley turned to Gus and squinted. He said, "Wait a minute. You've been telling us how this miracle of interplanetary colonization is almost within reach. Which way is it?"

Gus responded, "We will succeed. The question is how we handle the unexpected in a largely untested Mars environment. Please allow me to take another future excursion so this can all make sense. This is a very complex undertaking.

"We should not assume that we have discovered all of the problems that may emerge. We are capable of organizing and reasoning and making conceptual leaps that are ever more sophisticated as was common with the Apollo project. This is a new experience that is changing our perception of the natural world in many unanticipated ways. That's just the way new technology is implemented. Our intellect leads us. When President John F. Kennedy challenged us to go to the Moon and return by the end of the 1960s, no one dreamed of the many technical problems to be solved—much less their solutions.

"We do not foresee any insurmountable technical problems that would prevent us from colonizing Mars, but we do anticipate difficult problems. Furthermore, there will be human limitations and unanticipated events that will make the way easier in some ways and more difficult in others. We will make mistakes and we will make extraordinary discoveries. Our human personalities and insights will save the day again and again. Our direct involvement in challenging situations is important to our success and having hands-on people on Mars is at the core of reaching that transition. People do make a difference.

"Let me illustrate. Why has no one come up with this approach before? It appears so obvious now. The answer is that we were bound by our assumptions as to how the project should be accomplished. We are now thinking outside of that box, but there were some very real limitations. We absolutely had to find water and a way to create building materials from natural resources on Mars.

"Just two years ago, we conceived and tested a way to readily form stone building blocks if we could locate aluminum ore convenient to our base of operations, which is to say, near a source of water. We were very lucky to find water and aluminum ore together near the fossae site. That opened the door. Then we dropped the requirement to bring anything back from Mars for perhaps as much as a decade. Together these revealed an opportunity."

Sorley wanted to pin down some obvious tasks. "So how do you house these colonists for such a long time?"

"First, we have the experience with the International Space Station, which is a much tighter living space than the one we will construct on Mars. We have a lot of experience with growing plants and animals in confined spaces at the NASA Ames Research Center. Furthermore, one of the crewmen would be coming from Ames to manage that situation.

"Mars has a way of becoming enshrouded in dust for weeks or even months, which obstructs radio transmission. We need humans on Mars to control activities under such circumstances. Furthermore, humans can solve new problems very rapidly. While robots can learn within reason, they cannot deal with surprise situations well … and the inherent radio transmission time from Mars to Earth and back is quite long in those situations. There is no substitute for human adaptability and educated intellect. We would send an astronaut, a construction engineer, a geologist, and a life sciences expert as the first colonists. They would also be two men and two women."

Sorley asked, "What do we really gain by this besides a big adventure for us all?"

"First and foremost, we learn how to operate a technological society using only nuclear power produced locally … something we must learn if we are to evolve our society here with very limited fossil fuels and scarce mineral resources. We also gain a second home for humanity, a home that has almost as much land surface as we have here on Earth.

We learn an awful lot about the adaptability of people, animals, plants, and microbes.

"We learn from the geological experience of an entirely new world that once had seas and volcanoes and streams and extremes in elevation. We can obtain resources there which are already very scarce here. This will encourage colonization of the Moon, too.

"And finally, we take the first step outward from Earth carrying us where we do not know. Mankind becomes a space-faring species."

President Sorley said, "Dr. Bagnal, you certainly come prepared, but we still have the most enormous budgetary recessionary period since the Great Depression. I like what you have told me, but we just cannot undertake such an enormous project by ourselves. I'm going to assign the vice president to explore what might be done to bring in some partners. He will need expertise from our secretaries of state and treasury and my science advisor as well as you, Dr. Bagnal. I want a preliminary report soon. The vice president must also report back to me as to whether we can bring in some friendly nations to join us in such a project."

Dave needed to keep the lines of authority clear. He said, "Those nations must understand that they must keep this secret until we confirm feasibility of the project. NASA will direct the project in cooperation with the European Space Agency."

Gus said, "Mr. President, there is one more matter I need to raise in the utmost confidence." The entire room totally focused on Gus.

"We have been planning against many improbabilities that now seem more likely. We have technical sleepers among our probable first team to Mars. They do not know this even among themselves. I am going to speak in generalities.

"As one example, one of the team members has experience that does not appear in his records. Rob is cyber-brilliant. He was one of the team members that worked on the Iranian nuclear cyber attack a few years ago. We wanted him on Mars in the event that it should become a target of cyber attack—not just to be our construction guy.

"Rob also has experience with breeder reactors—reactors that produce plutonium from uranium-238 while using uranium-235 as a fuel.

"We have other similar technical expertise in the team. Joan has worked on uranium enrichment. She knows how to enrich any uranium we find so we can create fuel for our electrical energy generation power plants on Mars. We must find a way to ensure complete energy independence for the Mars project.

"Tim has been leading the design of the Base Development Module, which is the human and cargo section of the *Ares*. He also knows how to design and build a plasma rocket from his work with the Air Force. A plasma rocket could propel a rocket to Mars in a third of the time of a chemical rocket.

"Mary comes out of our Ames Research Laboratory. She has worked with artificial human gestation projects from in vitro fertilization to birth from an artificial uterus. She has brought our most advanced genetically modified food plants to Mars and knows how to further modify them if necessary.

"We will carry many examples of genetic fertilization to Mars. We have a good sampling of human eggs and sperm from exceptional donors that can be artificially implanted in our women. They can also be implanted in an artificial uterus so that real human babies can be artificially gestated on Mars—in part to ensure genetic diversity among our team. Growing children to adult age and education takes a long time though and their education would drain our staff.

"I have some more dramatic examples. I could go on, but this should be enough to show that we are approaching this challenge very proactively. We also have staff and access to brilliant people."

CHAPTER 7:

GENESIS CRATER

ATTEMPTING TO DIVERT the press from the obvious potential of human colonization was already going awry. Discovering water in quantity was stirring human ambitions. Finding minerals, particularly aluminum ore, capped the resource list, giving the public the impression that science fiction would now rapidly become science fact. On the positive side, this distracted humanity from its predilection with warfare. On the negative side, it raised expectations far beyond what was likely to be accomplished—and managing expectations had quickly become a major problem.

Four more teams were recruited to work on the project. One team would have credentials like Tim, another with credentials like Mary, and so forth. They were formed into a space and environmental team, a planetary geology team, a planetary engineering team, a planetary life sciences team, and an engineering construction team. Each would work under the direction of Tim, Rob, Joan, and Mary. They were fleshed out with support technicians and administrators.

Meanwhile, the Ares development team was reactivated and expanded with additional personnel to work with Tim on the crew components.

The teams would develop the ships and cargo for the next unmanned fleet. Other professionals were brought into the project to provide consultation in the areas of their specialties. The initial four were prime candidates to crew the first manned fleet.

Those who would crew the following fleets would be drawn from the additional specialists as the need for their personal presence became apparent. Professionals in nuclear engineering, electrical distribution systems, mining, medicine, veterinary medicine, transportation, chemical engineering, manufacturing, computer systems manufacturing and maintenance, and psychiatry were selected.

Once the first fleet of ships landed, the primary four teams would reform with one sci-geneer (scientist-engineer) of each type on each new controller team. The four controller teams on Earth could guide the work taking place on Mars in twenty-four-hour shifts.

Tim and his team had evaluated all of the *Ares* crew components. The model was no longer the old *Viking*—it was the *Ares* rocket carrying life support and cargo modules, but the old *Viking* names would be continued. Using a specific crew module would give away the prospect of a human landing.

The four *Thor* mission ships had to descend from orbital velocity using trailing atmospheric brakes, ending with a vertical profile and settling with no lateral motion on four shock-absorbing pads. Those last few meters of descent were much like flying a helicopter. Modern helicopters employed electronics that could pretty much set the craft down on a dime, assuming that there would be guidance references on the surface. The 2012 exploration mission had set the electronic guidance markers in place.

Tim's team had already assembled the humbee flight platform. Again, the LEM experience guided them. The rockets and their controls were already under fine-tune testing. The last of the components would be ready for assembly within six months.

The final six months before lift off of the unmanned *Thor* fleet were filled with innumerable simulations and tests. NASA and JPL were not new to space flight missions and readily adapted to the radical change.

The media and science fiction fan clubs were so hot about the possibility of a human going to Mars that Dave and Gus ran out of ways to avoid the subject. They sounded about as sincere as an ambitious politician avoiding reference to his political intentions.

Chapter 8:

Deceit

THE LEAD QUARTET of scientist-engineers was Tim, Joan, Rob, and Mary. It came as a shock when Tim and Joan were called in by Louise and informed just months before launch that recent assessments indicated that they would be in the first group.

Not only were personal technical abilities and teamwork being considered in assignment to the teams, the interpersonal relationships between the pairs were critical. They wanted to populate Mars with the children of the colonists.

Tim and Joan were outwardly bonding. Rob and Mary were very businesslike at work. Tim and Joan were very open about their feelings and that they were engaged. Rob and Mary were more matter of fact. However, the four would often slip away for group activities.

Meanwhile, another pair of engineer and life sciences professionals was making their personal relationship very obvious at work. The others were becoming annoyed by Ted and Candice's juvenile behavior. Candice took every opportunity to cozy up to Ted, pointing out their advanced amorous compatibility.

Tim's first thought was of their friends. He said, "Rob and Mary will be happy to be selected as well. We've been a team from the outset."

Louise frowned. "Actually, they will be on another team and they may even be dropped."

Tim and Joan were shocked.

Joan said, "That can't be! They're the most proficient at their professions!"

Louise grimaced. "We don't see any sign that they are compatible with each other. We can't take a chance of sending anyone to Mars who does not bond with his or her partner."

Joan said, "What do you mean? They just don't blather it about like some others."

Louise said, "The decision is made. We have on good authority that they might even be gay. We need breeding couples."

Tim rose to the occasion. He said, "Look. You may have some gossiper whispering in your ear, but you should have asked us."

"I did not want to offend you and your friendship."

"Mary and Rob have a very advanced relationship."

"I am very glad to hear this, which is to say they will not be dropped from the program."

"No, either they become our partners on the first team or you're likely to lose all four of us."

Tim looked Louise directly in her eyes and held her gaze, but Joan took the lead.

Joan asked, "Louise, do you have any idea how we spent yesterday afternoon?"

Louise did not reply.

"After working here all day, Rob and Mary joined Tim and me at his ranch. I live there with him. We had a couple of drinks by the pool and we all went for a swim. After cooling down for a bit, Mary took off her suit. I joined her in the nude and so did the guys."

Joan pulled down her skirt a bit to show a contraceptive patch.

"Mary has a patch just like this. After a little tantalizing horseplay, we climbed out of the pool with the guys giving clear physical evidence that our gamesmanship had full effect. We were all very stimulated. The guys chased us into our bedrooms, hardly taking time to dry off. Do you need to hear more?"

Louise was taken aback. She was blushing a bit and cleared her throat. "Do you do this often?"

Tim said, "Most weekends. It depends upon the girls and their cycles. Somewhat surprising, their cycles seem to have synchronized. On off weeks, we just sit and talk together with a rule that we never talk about work."

Louise was having trouble with her questioning. "Are you monogamous?"

Joan laughed. "Listen, if you saw what these guys pack, you wouldn't ask. We're very happy with our matching and would not change anything about it. We do want to build a swimming pool in one of the domed rooms at Craterwall."

Louise lowered her eyes. "We have already selected your partners."

"And I suppose they are Candice and Ted—your sources."

"How did you know?"

"Louise, it was a setup. They manipulated you!"

Louise was stunned. "My God!"

Tim saw his opportunity. "Now we have another condition. You have to drop those two. They would destroy the project. In fact, I was going to suggest that at our next meeting."

"They're waiting for you in Gus's office now."

"I suggest you call Gus and ask him to come here."

"Please leave for now. I won't let you down. I just need to talk with Gus in private."

Tim and Joan looked at each other and then left the room.

Louise called Gus on his private line. He appeared almost immediately at her door. Louise knew that she was exposing herself. Their conversation was brief. Both of them walked the few steps to Gus's office and Gus spoke to Candice.

"We have decided that Tim, Joan, Rob, and Mary will make up the first team. I would apologize to you except it seems you were the cause of our misinformation. After this, it would be impossibly awkward to have you around. I hope you understand this clearly."

Ted turned to Candice as they left the room. "What on earth did you do?"

There was no response. She was in tears.

Gus turned to Louise and said, "First, you did the right thing being up front with this. It's the only right way to handle it. But this will destroy the confidence between you and the team. It does leave us with a problem having you lead the Mars project. I need to work this out, but I need to swap you and Arnold. You go to the lunar project and bring him to be my operations head over here. I will take over the Mars project myself. This will take some fence-mending."

Gus spoke to his secretary and soon they were joined by Tim, Joan, Rob, and Mary.

Gus spoke, "First I want to apologize for not being in more direct contact with the Genesis Project. Joan is moving to the Lunar Project now and I will take direct charge of Mars. You're clearly the four we need to go on the first manned mission to Mars."

Everyone was all smiles.

Chapter 9:

Thor–Fission–
Tornado-Blitzen

S<small>PACE</small> <small>EXPLORATION</small> <small>WAS</small> being followed more closely by the news media than at any time since the Apollo missions. The launch of *Thor* would be spectacular. Enormous dual rockets would lift the much smaller ships from the launch pad and inject them into the elliptical orbit that takes it to its destination where another rocket burns each to the threshold of an elliptical interplanetary orbit. The most economical orbit is tangent to Mars orbit for transition into a landing approach profile. Because of the staged launches, the actual interplanetary orbits of the *Thor* rockets were adjusted to match the locations of the planets.

An elliptical orbit is a constant total energy orbit—the total of kinetic energy and potential energy in the sun's gravity field is the same everywhere in the orbit. That is why orbiting celestial bodies repeat traveling in their elliptical orbits. An interplanetary orbit is one where the rocket effectively coasts from one planet to another after injection by the rocket engine. The slowest velocity is upon reaching the outer planet's orbit when another rocket burn is needed to match the velocity of the rocket to that of the planet in its orbit.

*   *   *

Gus held the final news conference before liftoff. He said, "Stories about our undertaking have ranged from real science to the ludicrous. I feel compelled to make a statement that describes what those of us that lead the project see as our actual mission and our prospects.

"This plan is quite optimistic. Only if we are very fortunate shall we be able to proceed further with the excavation of working space within the crater wall. We will send more equipment, supplies, plants, and animals two years from now in the next Mars mission. I will take questions now."

Gus invited a reporter by name to speak.

"Dr. Hoover, you seem to have turned from optimism to pessimism. Why?"

"Mr. Thomas, what I have just outlined is all on the frontier. We do anticipate considerable success. We just want you to know that what looks like partial success to you may well be a great achievement.

"This undertaking is historic in scope. We ask you all to scale back your expectations. This is a very new experience for everyone involved. We will all learn a lot and adjust our next mission accordingly."

*   *   *

JPL was responsible for the transits to Mars and the landings. The first *Thor* ship performed flawlessly during transit, approach, and descent. Long, quiescent rockets again fired to slow the ship's progress, dropping it into an arching descent toward the surface. The propulsion rockets were separated and replaced by a trailing, thin-atmosphere braking device that slowed the lander enough to deploy the parachutes. A cluster of three drag parachutes deployed in the thin atmosphere until *Thor* approached the surface and intermittent plumes of rocket fire guided the ship to the surface so that the landing pads could skirt obstructions. The lander settled to the surface. The parachutes were cut loose from the ship only when radar sensed minimum descent velocity. The struts linked to the landing pads absorbed the load of the ship as the pads made contact.

There was no one onboard to send a reassuring message that all was well. However, the central control processor aboard *Thor* quickly assessed the condition of the landing and of all operating equipment onboard, then sent that status back to Earth by way of a geostationary communications satellite positioned above the Mars equator, due south of the landing site.

The ship was confirmed to have safely landed close to their target location near the Genesis Crater. Living creatures were among the cargo, making it the first permanent placement of life on another world.

The landing was watched remotely—but very carefully—by the four primary technical managers. They all lived their jobs, staying close to the informational and control devices managing activities on Mars.

They employed scripts for controlling expected activities and monitored inputs from others, developing and feeding them revisions as events moved along. Periodically they paused to actually assess what was happening and consider peripheral events that might suggest any revisions needed in the master plan.

Cameras onboard *Thor* photographed the surface south of the fossae from myriad angles and altitudes as the base development craft settled toward the planet. The images were captured by the geostationary Marscom satellite even before the landing and were instantaneously forwarded to Earth.

Tim was so intent upon the images and data that his hands seemed to move invisible controls. *Thor* slowed to a moving hover. When the limp drag parachutes whipped in the rocket blast, *Thor* found its footing and the jets calmed. The chutes settled rapidly to the ground just to the side of the craft. The billowing talcum-fine Mars dust settled slowly in the extremely thin atmosphere. *Thor*'s four shock-absorbing torsion mechanism legs spread its weight and balance widely from the body of the craft to the landing pads. The image was that of a spider. Rocket landings had returned to Mars. The spider had come to build a nest. With that, Tim seemed to relax in his seat.

The BDM consisted of the crew and biologicals compartment, a cargo module, the landing rockets, the landing shock absorbing frame, and the touchdown pads. The cargo module was attached beneath the crew compartment by explosive bolts. The cargo module could be lowered to the surface with extendable cables once the explosion bolts were released. The rockets were outboard of the cargo unit to provide landing maneuverability.

The cargo module was released and came to rest on the surface even before word of the successful landing had traversed the seven-minute journey to Earth. Two AI robots had already been released from the crew cabin and another from the cargo module. They had been given tasks to accomplish and immediately set about establishing position markers to guide their intelligence-directed equipment to critical locations about the landing zone.

Chapter 10:

Life-Giving Water

WORKING ON THE Mars surface offered some advantages over working on Earth. On Mars, there was no weather to speak of, no water collected on the surface, and few storms. The little wind that did develop was extremely light. There was no dust in the air except during seasonal dust storms. Beyond that, there were no plants or animals to inhibit or disrupt construction activities.

The explorer mission had drilled deep to find the ice aquifer about ten kilometers west of the crater. The sinking of a coaxial well pipe was the initial and most critical step. Joan and Rob monitored the loading of the water-drilling tower components on the hauler for their transport to the well site. Two telescoping legs of the tripod were set in place over the old drilling site and the robots used the third telescoping leg to leverage the structure into an erect position. All supports and reinforcements were locked into rigid position.

The solar panel from the exploration two years earlier was in good working order and activated. A new second panel was erected nearby and provided additional amperage in electrical parallel. Both units were designed to turn their photoelectric faces directly toward the sun as it moved across the sky. Virtually all days were sunny from the more distant solar orb that provided only four-tenths of the energy intensity that warmed Earth.

The ice aquifer had been located by the *Thor* exploration. Now the drill bit was guided down the original drill hole by a thin tapered lead, shaped like a hole-punch. The drill bit spun around the base of the guide, freeing long-dormant stone. The sections of the coaxial pipe behind the drill were fronted by a shield that would block any large stones from advancing into the pipe and becoming lodged. The inner and the outer tubes could be blown out, but the inner tube was particularly sensitive since it had to carry the small debris from the grinding to the surface. Each section of the pipe was designed with a spiral threaded wedge that would lock the two sections together.

A robot fit the first section of pipe into the drill bit section and locked it into place. A second pipe section was fit into the trailing end of the first and that was locked into the electrically powered drill mechanism at the top of the tripod.

Joan and her team confirmed that all was ready. Tim placed his hand on her shoulder in reassurance for just a moment. "You're right on target."

Joan gave her computer the command to commence. They waited fourteen minutes until the radio response. The command traversed the heavens. The drill mechanism began to turn and compressed Mars air from a pressurized reservoir tank was blown down the outer tube. The drill bit descended slowly into the ground, stopping when it had advanced enough to need a second section of pipe. The new section was locked into place.

Everyone was holding his or her breath as one critical action followed another, threading the drill ever deeper. Another pause. Another confirmation. Another section drilled downward toward its destiny. Ever so slowly, section after section was added until darkness fell upon their first day. With the setting sun, their solar power cells produced less electrical voltage and the AI shut down the drilling process.

It had been a short piece of a day that had taken them down only sixteen meters. Debris blown up the central tube was matched against the drillings from two years before. Everything was as expected. They had to wait a seemingly long time for sunrise and recommencing their effort to bring ancient water to the surface of Mars. Any sleep would be fitful.

Meanwhile, Rob monitored many other processes that had progressed. He placed the ship in near-dormant mode except for the environmental systems needed to maintain critical life support for its living cargo. Robots moved equipment and supplies from initial locations near the cargo pod to planned workspaces where they would be needed. By day's end, there were three distinctly active locations: the area around the ship, alongside the crater, and the drilling location. The earth movers powered by methane and oxygen had plowed out crude winding paths between these locations. The chemical reaction of methane burned with oxygen produces carbon dioxide and water, which is innocent enough to be released in the Mars atmosphere.

Rob directed the small backhoe to dig a notch into the side of the Genesis Crater's northwest quadrant and to begin construction of a temporary facility in the open, just outside the anticipated location for room NW9O (northwest room nine on the Outer Ring). That room would eventually be the permanent location for the distillation unit. For now, the impure water from the distant well would go to the outside unit.

The excavation operation was suspended at sunset. The floor of the pit was leveled. Marscrete powder was mixed with screened surface grit. An adjustable frame was set in the floor to outline the space for the distillation unit's base. Once the inside of the frame was filled with the Marscrete mix and packed tightly, the thin frame was removed.

Rob approved action by a robot to set an acetylene torch in position and fire it. The flame played upon the Marscrete mix for less than a minute, commencing the powerful heat-yielding thermite reaction. Enormous amounts of heat quickly spread within the mix and into the adjacent ground. The chemical reaction was unmistakable, but its consequences were measured for verification. Two robots lifted the distillation unit's steel base plate with deep anchor pins and sunk it flush with the molten Marscrete. Adjustments were made to ensure that the base was level—none too soon as the cold ground and the night air quickly cooled the mixture. The base was immediately anchored in a solid stone foundation.

Similarly, another pit was dug to the opposite side of the well. Another attachment plate was rigidly set in a thermite block. A combination heater and air compressor anchored to the plate would be used to force hot air down the outer section of the well's pipe.

*   *   *

Mary and her team resumed the well drilling operation at sunrise. Torque was applied to the shaft, but had no effect. The power was increased, but the shaft still did not spin. The torque was higher than it had been during the first day. The shaft was designed to carry the load. Then the shaft abruptly spun wildly. A safety cut in and stopped the drilling mechanism.

Joan let out a gasp. She knew what that likely meant. The drill bit must have jammed overnight. It was surprising since the bit had been hot and with cooling it should have contracted and become loose. The next step was to analyze the last debris brought up the day before. Examination revealed that the debris had included a little moisture that most certainly froze tight overnight. The free moving upper portion of the shaft was removed, revealing a spiral twist fracture about nine meters down the shaft.

Joan had a determined look. She said, "That should not have happened— even if it froze into place. We have a spare section of the shaft here. We knew normal steel would lose strength in the cold. The shaft was made of a special alloy to keep most of its strength in the bitter cold, enough to have withstood that torque without failure. The manufacturer certified the alloyed steel. I'm going to test the sample to see if he gave us what he said. Meanwhile, I'm going to start circulating hot air down the shaft."

Gus waited for his other professionals to react.

Rob said, "We really need to retrieve the bit. I have something of a long shot. We sent along some epoxy. We can't repair the shaft with epoxy to give it enough strength to proceed. Maybe we can give the shaft enough strength to withdraw it with counter rotation."

Joan agreed, so after the heating, they had a robot mix and place a small amount of epoxy on the break and lower it until it made contact with the embedded section. The mixed epoxy generated heat as it bonded. They waited more than enough time for the bond to fix and tried the counter-rotation withdrawal. Since the shaft disconnected above the break, a small amount of epoxy was applied to that joint. It was lowered until it screwed together with the epoxy and was allowed to set.

The procedure was repeated until a joint below the break released and they could withdraw the broken section. The entire drilling pipe was removed from the well and the damaged section was replaced. The entire process was tedious. Since it was not preprogrammed for the robots, every step took a lot of time.

Meanwhile, the results were back from testing the spare section of pipe at Earth Control. The cold test on Earth caused failure again at the same torque and the metal showed that the manufacturer had misrepresented the alloying of the steel.

Gus and Dave were both livid. Here they were at the most dramatic moment in their project. Dave asked, "Who was the contractor?"

Gus replied, "They're Chinese. They sent test results with their shipment. We specified that all tests were explicitly to be done at cryogenic temperatures. I'll bet they did the test at room temperature."

Dave was still furious. "Damn foreign relations. We will take no more shipments from the Orient. There is too much danger of misunderstanding and cutting corners!"

For want of a nail, a shoe was lost ... until for want of a battle, a war was lost. Everything was at risk because a manufacturer had cut corners and lied about its work.

Four days later, they were ready to begin again with new shafts. After six more days of careful drilling in heated workspace, they still had not reached the top of the frozen aquifer.

Meanwhile, *Fission*—the next ship due to land—was experiencing its own problems.

The components of the nuclear power plant would arrive on the other three ships. *Fission* carried the nuclear furnace—the 5–6 percent radioactive isotope uranium that would heat the primary circulation of water into steam. *Fission* approached Mars ten days after *Thor*, but was following a slightly misdirected descent.

If it landed from that path, the ship would drop into the nearest fossa and most likely crash—and be unreachable on the ground.

Tim and the JPL team made some quick adjustments and the JPL team executed an unplanned retrorocket firing in the hope of at least keeping *Fission* on the upper plateau. Furthermore, instructions for the descent were altered so that *Fission* would execute a radical landing maneuver. No one knew if this would be enough. There was no more time for further instructions as *Fission* entered the radio delay blackout period.

The entire team was anxious; the loss of *Fission*'s cargo would virtually eliminate any hope of underground excavation. Abundant Marscrete in particular was essential to reinforcing the arched ceilings that could hold everything underground in its place. Electrical power was critical to manufacturing the thermite mix.

The ship came in high and fast. The parachutes were deployed as soon as the ship penetrated the air that was dense enough to open their canopies. Landing rockets then commenced firing, slowing the ship's forward movement rather than saving rocket fuel to soften its landing. The ship turned to a near-vertical descent near the rim of the fossa with the rockets still burning. Virtually all of the fuel was spent as the ship reached low altitude and then the fuel gave out. The ship dropped a good ten meters to the surface, the pads struck boulders, the struts collapsed, and *Fission* rolled heavily to its side. It was barely ten meters short of the fossa and still on the plateau, but most likely with serious damage to its cargo.

The groan about the control room was audible. One of the AI robots on the ground, Sam, had already boarded a front loader and was on the way to the crash site. Two others had reached the well site and were loading the tripod support, used for well drilling, onto the rover. The only positive aspect to the crash was that—without any methane and oxygen rocket fuel remaining—there hadn't been any fire.

JPL had been in charge until the landing, such as it might be. Tim was responsible now. He had to rescue whatever was possible.

Sam seemed almost human in its reaction. It climbed up on the ship toward the crew hatch, wrenched it open, and checked the life-support system. That had failed. The rabbits inside were in an enclosure that automatically sealed itself to prevent loss of critical air pressure. The remaining air pressure inside the crew enclosure was quite low. Sam released the clamps that held the enclosure in place and tugged it out of the hatch with superhuman strength, carrying it down to the ground. Sam then shoved the enclosure into the crew cab of the hauler, connected hoses for oxygen and nitrogen to the rabbit enclosure, and opened the valves. Long, sealed breathable air containers explosively expelled a balanced atmosphere through a regulator.

Sam then climbed inside, sealed the cab, and inspected the condition of the rabbits. He reported that they were clearly the worse for wear, but whether from lack of atmosphere or internal injuries was not obvious. Sam issued a command to the hauler, which immediately trundled off in the direction of *Thor*, which had the only working life-support system then on Mars.

The crash site was ten rough kilometers north of *Thor*. Sam carefully inspected each of the rabbits, finding twelve still alive. He had already adjusted the temperature to normal. The image of Sam and the rabbits as seen through its camera eyes was touching. No human could have cared for the animals more compassionately.

Meanwhile Tom and Chuck, the other two robots, transported the well drilling tripod frame to the crash site. Using the frame, they raised the wrecked ship to something resembling an upright position and propped it up with boulders. Tim stepped in and, using jumper cables, he ordered Sam to fire the explosive bolts and manually lower the cargo pod.

The JPL team and the quartet were amazed at the ability of the robots to adapt to such an unexpected situation. Sally Thompson, their JPL teammate, was beaming. The robots were her creation. They had evolved over the past decade from largely preprogrammed workers into remarkable helpmates.

Sam secured the rabbits in the *Thor* cabin, making them as comfortable as possible. It then returned to the crash site. Tom and Chuck were busily unloading the cargo onto the ground in strategically located stacks. Most of the shock-absorbent cargo cases seemed none the worse for their impact. Finally, they reached the nuclear furnace and, using the tripod, positioned it on a tarpaulin where they could inspect it for damage. Upon first examination, it appeared okay, but then the sharp magnifying eyes of the robots in ultraviolet light discovered two hairline cracks.

Repairing the cracks was questionable. Their only hope was to somehow work fine powdered pure thermite mix into the cracks and fire the mix. That was Tim's decision. Thermite had been used to weld iron for more than a century. Hypodermic syringes in medical packs were available to use on the animals should that become necessary. The first test of their improvised solution was conducted by Sam inside the cabin of a hauler with atmospheric pressure. A thermite mixture was placed between two smooth, flat blocks and ground as finely as possible. Tim placed a needle on the end of a syringe, then put some thermite inside, and inserted the plunger to close the syringe's cavity. Sam pressed the plunger down into the syringe and, marvel of marvels, the powder blew out of the needle. The test had worked.

The next step was to load the syringe inside of the cabin to gain air as the propellant. Sam used his magnifying eyes to find the cracks, insert the needle in the cracks, and propel the powder deep inside. With luck still on their side, the powder penetrated the cracks. An acetylene torch was used to heat the critical location of the cracks, raising the thermite's temperature to the point of ignition. The torch had heated the metal, causing it to expand and squeeze the cracks and thermite tightly, closing the cracks. The ignited thermite produced intense heat to fuse the metal together. Afterward, there was no remaining indication of the cracks. There was no way to know if the repair had been truly successful.

The furnace was moved to Craterwall just outside of where it would be permanently housed. That entire operation had taken three days. The furnace would wait until events caught up while a pit was excavated for its temporary outside site.

The well was still critical, but had yet to yield any water. The tripod was moved and reassembled. Drilling of the well resumed its slow progress, continuing for four more days until it reached the top of the ice aquifer. Joan watched the residue readings as the drill penetrated the ice layer. For three more days, the drilling continued until it reached the bottom of the aquifer where it was stopped. The location for the intake would ensure that the melting water would flow downward until it reached the well's intake at the bottom of the aquifer.

Tornado, the third ship, approached for its landing. It carried the steam turbine that would convert the steam's energy into mechanical rotation of a shaft. Fortunately, *Tornado* was near centerline for its planned approach. Confirmation came from its instruments that it was on course for the planned landing. This ship behaved well, landing much like *Thor*, safely and near its hoped-for landing site. The JPL control team and the Mars Colony team were immensely relieved. Meanwhile, there had been only a brief pause to organize the equipment for assembly operations as *Tornado* executed its approach and landing.

The energy for these many well- and water-related activities came directly from solar panels and indirectly through fuel cells that were charged by the solar panels during daytime. The distillation unit and the air compressor–heater unit were set on their bases and bolted down. The wellhead was connected with pipes that would fill the transport containers. All of the equipment—including electrical and plumbing connections—was tested for tightness and operational function. They would not resume activity until dawn to ensure a maximum period of sunlight during the first day of ice melting and pumping operations.

Tim, Mary, Joan, and Rob with their backup teammates were tense but confident. There was little sleep while they waited for Mars sunrise, which came at 7:38 p.m. The worldwide news media was poised for this momentous occasion. It was the first of many critical tasks that would receive such scrutiny.

Joan initiated the operation by signaling the *Thor* AI controller to begin cold air circulation down into the well and back to the surface. It continued for a half hour. The heater element slowly warmed the air. The outer casing was warmed first, then the circulation was reversed and the inner tube was warmed. The last thing they wanted was to melt ice and send humidity up the inner tube while it was still below freezing. Heat from the descending air warmed the casing and began melting the aquifer's ice along the pipe.

Heat was reaching the depths of Mars for the first time in over four billion years. Only when all was warm could the circulation be reversed and the compressed air raise the liquid to the surface. It was not sparkling spring water, but it was the stuff from which everything else could be derived to make Mars a home for humanity.

Having water was the ultimate cause for celebration throughout the NASA family—and particularly for the four people who had carried mankind on its first real step toward living on another world. The global media and those they touched sensed that a page had been turned. With water, people would somehow find a way to turn the rest of the human dream into reality. In an inspirational way, this was the same as the mural on the ceiling of the Sistine Chapel.

CHAPTER 11:

SURVIVAL ESSENTIALS

THE FIRST LIQUID to reach the surface was spilled out into a prepared depression. No one wanted any surprises. The liquid proved to be rather cloudy in appearance, indicating that many suspended contaminants were present. The distillation unit would evaporate the murky groundwater into steam and condense that into pure, clear water. The residue would be collected into a sludge chamber that was periodically flushed into a collection unit for its potentially valuable chemicals.

The first use for the pure water was to support living things: drinking water for the animals, swimming water for the fish, and water for the various plants.

Chemically, H_2O was first used to produce hydrogen and oxygen gases. Tim oversaw this operation. The facilities to do this had also been dug into the wall of the Genesis Crater at NW70 (northwest room seven on the Outer Ring of the crater wall). The pure water was then electrically dissociated into hydrogen and oxygen. The gases were compressed and stored in high-pressure cylinders.

The other natural resource that would prove most useful was the carbon dioxide that dominated the content of the thin atmosphere. The atmospheric gas was compressed, and then refrigerated until the carbon dioxide froze, becoming dry ice or pure carbon dioxide. The remainder was still a gas, mainly nitrogen and oxygen, which was returned to high-pressure cylinders.

Rob was responsible for the atmospheric gases facility while Joan saw to the next process. The pure carbon dioxide was warmed to become a gas. She mixed the carbon dioxide with pure hydrogen gas to chemically produce methane.

By this means, fuel was produced for internal-combustion-powered equipment, again using only natural resources from Mars. It was oxidized in the engines using the atmospheric residue of nitrogen and oxygen. The base thus became able to power the internal combustion engines of the backhoe and the hauler as well as other smaller equipment.

Pure water had been produced, but that was twenty-five days after *Thor's* landing. Thirty-five days after the landing, pure water, hydrogen gas, oxygen gas, carbon dioxide gas, mixed oxygen-nitrogen gas, and methane gas were all being produced and the young base possessed limited self sufficiency with fuel and some critical natural resources to mix a breathable life-support atmosphere.

Only the well water retrieval and the water distillation were conducted at the wellhead. All of the other operations were set alongside the wall of the Genesis Crater, but they would be moved inside once interior space was bored out and lined as domed chambers.

Rob and his colleagues were deep in their first serious construction project, digging a pit for the nuclear furnace, a steam furnace, and one of two electricity generators alongside the Craterwall site. The team was living and breathing the first serious attempt by people to construct a facility on another planet. All previous facilities had been self-contained. In this case, the components were being installed separately and connected to work in unison. The indigenous materials were being converted into new materials and formed just as they were processed on Earth.

A lot depended on the artificial intelligence robots developed by NASA and Cal Tech scientist-engineers. The group became very close in their thinking and resolve.

In the mid 1950s, the US Army Engineers had constructed a portable nuclear power generator to power the Radar units on the arctic Distant Early Warning line. The Radar units could warn of approaching Soviet bombers from over the North Pole. Those power generation units could be air-dropped into distant locations and burrowed into the frozen tundra. The nuclear generators taken to Mars were a modern and more compact version of this historical concept. The approach had been proven repeatedly using aircraft and early electrical equipment.[2]

The components would provide the power to begin tunneling into the crater wall and boring out the first rooms to house the equipment. It would also be used to produce aluminum powder that would be mixed with magnetically enriched iron oxide in the surface dust. This combination

---

2   Note: In November 1956, the author was a young lieutenant at the US Army Engineer School at Fort Belvoir, VA. At that time, he delivered briefings on this equipment and its intended use.

resulted in Marscrete powder. Fired Marscrete blocks would reinforce the underground tunnel and dome surfaces.

*   *   *

The team members were discussing the ramification of their work.

Rob said, "Do you realize that just one of our decisions can influence the future of all mankind? Whether we succeed or fail is like monkeying with the course of history. If our project works, then our people will carry the ball for quite a while. If we fail, then someone else will take the initiative. Consider that we may not have found a way to fix the break in the drill with epoxy. Consider what happens if one of the critical elements in the power generation scenario fails or if one of the landers just hits the ground hard. We are running behind schedule now. What if we get into a fix on catching up?"

Joan shook her head and said, "This kind of thing happens all of the time with airplane crashes."

Mary frowned and said, "Look, we could have a big dust storm when we're landing. Those things happen. Yes, our work's time critical and single-threaded at the moment. Still, we do our best and shrug it off. Besides, this could go positive, like making a big discovery of some kind.

Tim said, "I know this work is wearing, but I'm really enjoying it. Who else gets to live history with a bunch of weird characters like you?"

They all laughed. They were enjoying the adventure.

*   *   *

The generator would produce their first serious power on Mars. The outside construction concept was basic and temporary. It would sink the assembled equipment into the surface just deep enough for the walls of the pit to shield surface activity from nuclear radiation.

The pit was deep enough so that the radiation emitted from the furnace was either absorbed by the walls of the pit or escaped into the sky. There would be no fancy radiation shielding.

The pit contained the reactor with a primary steam pipe loop passing through a heat exchanger. A secondary steam pipe loop passed through the heat exchanger and out of the furnace chamber to the adjacent electrical generation chamber powering the steam turbine. The steam pipe then looped back to the heat exchanger. The steam in the primary loop would be exposed to radiation in the reactor. The steam in the secondary loop was not.

Upon assembly of the two loops of pipe high pressure was used to test its integrity. The pipe making up each of the loops was connected, fresh water pumped into the pipes, and a high pressure pump used to raise the pressure above that expected from the steam. This was called a high pressure hydro test. They did not want to have a break in a radioactive circuit with live steam.

The steam turbine was the largest piece of equipment being brought to Mars. The generator pit was excavated and the turbine was carried from *Tornado*'s landing site to Craterwall by the combined effort of the two haulers. It was hauled down an incline to the floor of the pit. The chamber was also shored in stone walls filling most of the excavated space. There was space only for the turbine's base which was set into Marscrete.

The secondary steam line was completed from the furnace pit. With all of the steam systems in place, the nuclear fuel rods were put in place inside of the reactor furnace and distilled water was placed in the reservoir to feed the steam lines and thus the reactor. Only low-pressure steam was produced in the primary loop to warm the reactor space. Nuclear-generated steam had been produced on Mars.

Blitzen carried the two large generators and most of the animals. The approach for landing had been a bit wide and to the west toward rough terrain, but everyone was optimistic. Radio contact with the emerging base failed as *Blitzen* entered the atmosphere. All they could do was wait. Apparently the AI guidance was as able to handle emergencies as had the robot, Sam. The ship was easily seen as it cleared the horizon with its fully deployed parachutes. Then, still far out, the landing rockets fired and the ship seemed to be coming almost straight at the landing site. No one could imagine how that had been accomplished. The ship arched down to a perfect landing just one kilometer from *Thor*. That was the equivalent of a bull's eye, considering that it had been launched at Cape Kennedy.

Everyone celebrated. The robots unloaded the cargo immediately upon rocket shutdown. The cargo was found secured in place and the life support had brought all of the animals through in good order.

No one had expected four new design ships to bring everything to the Genesis Crater without any mishaps. The media went wild—as did people the world over.

But the game was not over.

Only one generator was moved to the power generation pit alongside the steam turbine. With the last Marscrete powder carried to Mars, they laid a base for one generator. The generator was set on the base and connected to one end of the two powered shafts that protruded from either end of the steam turbine. The construction team was again playing with no room for mistakes.

The electrical connections were made to the control panels that would direct power to the other machines, most importantly to the aluminum smelting chamber, but that was another story. In the meantime, the new electrical power was connected to replace the power from the solar panels at the Craterwall facilities.

Chapter 12:

Aluminum Acquisition

THE ALUMINUM MINE was extracting ore found some twenty-five kilometers away from the crater. Producing aluminum from ore was a two-stage operation. Refining would chemically produce alumina—aluminum oxide—from the ore. That was best done at the mine site if possible, thus reducing the weight and volume of material to be transported.

But providing electrical power and heat for the chemical processing at the mine site was not feasible. Consequently, the refining would have to be done on the open surface at the crater near the electrical and heat sources. Tanks to contain the solution that would dissolve the alumina from the rest of the ore would need to be constructed.

Then the alumina needed to be washed at Craterwall (SW9O) and loaded into smelting pots where the electrical Hall Process would produce pure aluminum powder.

Thermite powder was mixed at Craterwall (SW7O), further mixed with aggregate to make Marscrete, molded into structural shapes, and fired to produce stone blocks of various shapes. The mixed Marscrete would also be spread on prepared floors of pits and fired there to produce smooth stone floors.

Meanwhile, the second water well was drilled at a proven site to augment the water supply needed for refining. The well would duplicate the one drilled ten kilometers west of the Genesis Crater. Sinking the well would use the last of the pipes and the last remaining solar generator panels.

Haul Road ran from the aluminum mine operation. The refining operation near the crater was set up as the power plant was being placed into operation. Trucks hauling ore in to the refining site would use their dozer blades to improve the winding road on their outbound trip. Both loaders were used to set-up the mine and then one was positioned to load the trucks while the other assisted in setting up the refining facility.

Relatively small stone panels of Marscrete were fired to make the sides of the tanks that would be used for chemical processing. The panels were then fused together at corners to create stone-walled boxes. The tanks were only three meters by four meters by two meters deep, loaded with solution and ore, and then pressure-sealed and heated. [3]

Smelting employed the electrolytic Bayer process to reduce the alumina into pure aluminum. The alumina is dissolved in a cryolite solution. Cryolite (Na_3AlF_6) is derived from fluorite, which is a rather rare mineral, but fortunately that is not used up in the process. Direct current electricity is passed through that solution and precipitates the aluminum, leaving the cryolyte. The process is much more complicated than that simple description and requires very large amounts of electricity. For that reason, it was located as close as possible to the nuclear power generator at the crater and would consume the larger share of the produced electricity.

Getting the refining and smelting operation working was extremely tedious. It took four more weeks to get the refining process into operation producing alumina.

The first load of bauxite was delivered to Craterwall very slowly over

---

3   Note: Alumina is a white granular material, a little finer than table salt, and is properly called aluminum oxide. The Bayer refining process used by alumina refineries worldwide involves four steps: digestion, clarification, precipitation, and calcination. To turn bauxite into alumina, grind the ore and mix it with lime and caustic soda, pump this mix into high-pressure containers, and heat it. The aluminum oxide is dissolved by the caustic soda, then precipitated out of this solution, washed, and heated to drive off water. What's left is the white powder called alumina, which is transformed into aluminum metal in the smelting process. —www.alcoa.com

rugged terrain. There it was loaded into the refining tanks. The precipitated alumina was washed and the wash water was dumped into a nearby depression where the water evaporated and the residue was stored nearby.

The washed alumina was dumped into the precious cryolite solution in the smelter tanks. Power was turned on to the tanks and, seemingly by miracle, the metallic aluminum powder began depositing at the anode. The lengthy, tedious process had been practiced many times by the Genesis Team, but seeing the results on Mars was real cause for celebration.

The aluminum powder was the critical material for all that would follow. It was produced initially in very small quantities. The first action using the new powder was to enlarge the tubs used for refining as well as construction of a second smelter pot. With accelerated production of aluminum powder, virtually everything they had planned was becoming reality.

CHAPTER 13:

INTERIOR CONSTRUCTION

ROB WAS THE central figure for construction of the Craterwall interior. The plan for their new habitat had been put in place much earlier. Virtually all of their equipment and AI robots would be employed.

The first Marscrete blocks were used to construct the western underground Craterwall entrance. The natural crater wall began with a relatively low surface slope. That part of the passage was open to the sky and gradually dug into the crater as it rose high enough for the passageway to require an arched ceiling. Still, the passageway could not enclose the atmosphere until the weight of the crater upon the passageway ceiling was enough to compensate for the interior pressure.

A pressure-tight doorway would eventually close the passage against dust storms. The passageway penetrated deeper until the overburden became heavy enough to counter the anticipated internal atmospheric pressure. A pair of sealable doors would be installed to enclose an entrance chamber that could be pressurized with an oxygen-rich atmosphere. Outside was a native Mars atmosphere, but inside would be the breathable interior atmosphere.

Moving inward from the pressure chamber was a four way intersection linked right and left to the Outer Ring and continuing inward as the West Radial passageway. The next project was building the Outer Ring southward with construction of three domed chambers on the outer side of the ring. The first chamber would be used initially for the living things that survived thanks to the *Thor* ships' life-support systems. The second chamber would house the nuclear furnace. The third would house the steam generator and the two electrical generators.

Rob guided the completion of the interior nuclear furnace chamber and the electrical generation chamber immediately, but the broken drill bit for the well had put them way behind schedule and early winter was approaching. Everything was crowding them now. All of the outside processes needed to be installed inside before the oncoming bitter cold would freeze the water flowing from the wells and that would prevent all of the outside processes. It urgently needed to be inside and working properly.

Rob was very restless. Mary shared his concern. They were snuggling in the comfort of Earth weather. Mary had tried to divert him, but their embrace was only a temporary respite.

Mary gave him a playful after-sex nibble. "Come on. We are doing all we can and we are still okay. I know it's close, but you can get it all working."

"Gus is not so sure, and neither am I. We can't afford any more delays."

"Look, everyone is working day and night and it's going smoothly now."

*   *   *

Summers on Mars were like a severe northern mid-latitude Earth winter without snow. The winter was like an Antarctic polar winter. Polar-generated dust storms tended to form in the spring.

Getting the generator working right away meant that their power source could be protected beneath the ground—and the heat from the spent steam could be used to heat the interior as they extended the construction.

Every effort was made to get the first chamber and the passageways durably constructed. The domed chambers were tricky since a central column of dirt was left in place to support the ceiling as the Marscrete block walls and dome were raised ever higher. It took weeks to get the process working right because of the low Mars gravity.

Constructing the first domed chamber was a learning experience. The second and third chambers south of the West Radial intersection came much faster. They were all tested thoroughly for solid and airtight construction. The chambers had paved floors and bases with plates for the heavy equipment. All of the outside equipment was brought into the new space.

The inside plumbing was soon in place. Then the outside system was turned off and all electrical generation reverted to the solar panels.

Most challenging was maintaining minimum solar electrical power while installing the nuclear power plant components inside of Craterwall. The entire outside power generation system was then turned off and allowed to cool.

The first step in making the move was disconnecting and moving the original radioactive nuclear furnace from its outside pit and placing it onto a new base inside the second chamber. Since this was accomplished by robots, there was no danger from residual radioactivity. A hydro test was made with each loop before moving on to steam. The nuclear fuel rods were slowly advanced into the furnace, introducing steam into the primary steam line and making it possible to pressure test with live steam.

When the heat exchanger came up to operating temperature, water from its reservoir was heated to steam and allowed to flow through the steam turbine bypass and through the entire secondary steam line.

The turbine steam bypass was closed. It directed the steam to pass through the turbine, and for the spent steam to reach the network of heating steam lines and loop back to the input steam line for the heat exchanger. That meant heating of the active Craterwall facilities was serving as a cooling tower for the spent steam.

The original outside generator was then moved inside onto its new base and connected to the other end of the turbine's shaft. All of the other equipment requiring electrical power was added to the electrical load through the power distribution system until the inside system was drawing power. The final step was to put the completed system through every emergency procedure imaginable for managing every predictable problem.

The passageways had their own complications. They doubled as carriers of the utility distribution and collection systems—along the side of the floors and the ceilings. Electricity was carried along the ceilings as were communication systems and atmospheric circulation ducts like air conditioning systems. Water, sanitation, recyclable water, fresh water, oxygen, and nitrogen feeds were within baseboard recesses along the floor.

An outside power distribution system was needed to reach all of the exposed electrical equipment. The first step was to manufacture pole like, three-meter columns of Marscrete. Each pole had a hole down into the top that would accept an aluminum rod with glass power line insulators.

The next step was to begin extruding aluminum alloy wires to be fed through those poles. The wires would carry relatively high-voltage power along the outside of the Craterwall to the working operations and then on to the well. Higher-voltage power would eventually be transmitted by a string of stone poles out to the mine. With this power available, the refinery could be moved to the mine site.

Improvements in the twenty-five-kilometer Haul Road were being deferred as the controllers pursued more pressing construction. Construction of the Outer Ring passageway north from the West Radial intersection led to passageway extension construction reaching all of the chambers needed to bring the early physical and chemical processing inside.

Chapter 14:
Animal and Plant Habitat

ROB INSTALLED THE life-support systems only along the southern section of the Outer Ring within the three first chambers—for the biologicals, for the nuclear furnace, and for the steam turbine and two electricity generators. All of the interior heat for these chambers was drawn directly from the spent steam, using air circulation ducts installed along the ceilings.

Rob needed to balance the interior atmosphere of oxygen, nitrogen, and carbon dioxide at 12 psi. Marscrete hydroponic structures were constructed and temporarily installed in the biologicals area.

Mary controlled the transfer of the plants and animals from the *Thor* ships and into Southwest Room 1 on the Outer Ring, that first chamber. The contents were set about on the floor in the middle of the chamber.

Mary moved the animals next. Most of them sensed elbow room for the first time, giving them their first Mars home. The remaining rabbits were kept in cages and the few remaining fish were released in runs of flowing aerated water.

Rob extended the West Radial eastward to the Inner Ring then northward along that new ring. He added the habitat chambers along this short section of the Inner Ring. Meanwhile, the original chamber located first on the Outer Ring was still the only completed habitat space.

Bitterly cold weather was upon them. The mining operation, the refining operation, the mining well operation, and the original well operation had been prepared for the long winter and shut down. Blocks of pure ice were collected in four ice-house cells to provide a reserve of water in case of emergencies.

The heavily used equipment was wearing out and moved into a repair area. There it could be repaired to the extent that their few remaining spare parts would allow. The season was now six months past equinox and just three months before the second fleet of *Thor* ships would be departing for Mars.

With activity on Mars mothballed for the winter, there was little construction to be done. Attention focused on the plants and animals. Four cells were to be constructed on the Inner Corridor for growing soil-rooted plants, but those would not be ready for months.

A few rabbits had survived and were multiplying. The tilapia had produced their eggs, which by the method of their species were sucked into their parent's mouths to hatch. Suitable vegetation was placed in the running water troughs to feed the adult tilapia.

Decades earlier, there had been a question as to whether the dust on Mars would support growing vegetation. *Viking* had tested the surface dust for chemical composition. Equivalent dust was formed on Earth and all sorts of tests were made to see if it would support plant life. Unfortunately that was not to be. Dust from the moon would support plant life, but dust from Mars would not.

So it became a human enterprise to somehow formulate a soil on Mars that would do the job. The stems, roots, and leaves of hydroponic-grown plants were recycled into mulch, upon which termites and subsequently earthworms would dine. Then bacteria were added to finish the job. The processed mulch was mixed with the animal waste from the chickens and rabbits and the white water from the fish tanks. Ever so gradually, a bio-cycle was becoming self perpetuating.

At this point, the greater limitation on development was in equipment to take advantage of the new resources and space. There were only the two haulers and the two loader/backhoes as mobile equipment. There was one mole to excavate along corridors. There were only seven AI robots—and they were quickly wearing out.

The biggest missing aspect on this robot-driven mission was their very limited ability to manufacture anything. The only success was to extrude aluminum wire and to produce aluminum stanchion components to carry the electric power lines.

A pair of additional chambers was completed on the Outer Ring for which Marscrete was mixed and spread upon the floors and then fired. It would house the incoming nuclear furnace and the electricity generators. One additional chamber was constructed across from the original habitat chamber; it was sealed and environmental services were added.

Three months before the launch of the *Venture* fleet, the Craterwall facility had become a promising habitat for bigger things. Animals and plants had demonstrated that they could live on Mars. Their kind would also provide the food needed to feed the prospective colonists.

The biological space was supported with water, oxygen, nitrogen, atmospheric humidity, temperature control (mainly heating), lighting in various degrees, and electrical power. Construction was one thing. Managing a limited-space ecosystem was much more complicated and sensitive than anyone had anticipated.

Perhaps oddly, Mary saw her small ecosystem as animals living in a world populated by plants, so the plants were introduced first. Just three months before the planned launch of the first humans to Mars, the living things were comfortably living inside of the Craterwall Base. All the members of the team were proud of this accomplishment, but it was up to Gus to take the project to the next level—convincing President Sorley to give the go-ahead.

*   *   *

Gus and the colonist team were called to the White House to brief the president and hopefully to gain his blessing for launching the human adventure to reach Mars in 2016. President Sorley did not want his first people reaching across interplanetary space to die in the process.

The greetings in the crowded Cabinet Room were brief and subdued.

"Dr. Hoover, you have some convincing to do."

"Yes, Mr. President. I will not deny that we are four months behind our optimistic schedule, but we are still within our minimum guidelines, if just barely.

"We have constructed our basic passageways and have built six domed habitat rooms—two for the power generation system, one for the biohabitat, and three others as prospective work areas. The rooms are all operating with a viable ecosystem that is now supporting all of the remaining plants and animals that we sent. It has been operating for only a short while, but all of the acquiring of natural resources and their conversion to our needs to support that habitat and to construct passageways and rooms has been developing for the past fifteen months. Our technical team at JPL and the equipment sent to Mars have proven themselves remarkably well."

"So you are asking for my approval." It was more of a doubtful question than a statement.

"Only after we explain our current situation and what we hope to accomplish, sir."

The president nodded.

Gus said, "The prospective individual launch dates for the four ships are fixed by the position and movement of the planets. We need to send all four ships or none. We either go or we don't. If we miss the launch dates, we will have to wait two years until the next biannual opposition with Mars and our equipment is seriously wearing out. We still have much work to do on Mars before we can consider launching.

"All of the remaining equipment needed to supply life support must be moved from outside locations and into additional rooms that we are now constructing. That equipment must be tested and all of it must work in the new locations within the underground base. That effort includes construction of eight more rooms, extending the passageways, and installing utilities throughout. Furthermore, the current electricity generation and the electrical and heat distribution systems must be working reliably. The current and expanded environmental support systems must work reliably for the next nine months. The living things must thrive. The humans will need the same life-support that the animals use—this is our way of testing the team's working environment within the base."

"Dr. Hoover, am I correct that you believe you have everything on Mars now working as it will be needed by the four people who are here with you today?"

The comment reminded everyone that the prospective Mars colonists were in the room with them.

"Yes, the bare minimum of animals, plants, equipment, and base facilities that they would need are already on Mars. They would be supplemented by the new animals and plants that we are sending. We will also be providing more equipment to protect against malfunction, and additional equipment to give the crew many more capabilities ... but what is there now would see them through."

President Sorley wanted to be sure that he understood what was being proposed. "Let's check a few points. What happens if the electrical system fails? That includes the nuclear furnace, the steam turbine, and the generators. Do you expect to prove the reliability of the power, heat, and atmospheric gases distribution systems in the next three months?"

Rob said, "Mr. President, that system has worked reliably since it was assembled on Mars. The nuclear furnace has dual functionality and many dual systems. It is internally backed up. Furthermore, it has no external moving parts except pumps and those are also dual. The heat exchanger between the primary and secondary steam loops is a dual system. The steam turbine has dual plumbing, but there is no provision for the turbine itself. That is why we are sending a second steam turbine on the first ship that will be installed in the room adjacent to the nuclear furnace, the room now used as a biofacility. The current turbine serves two generators so they are backed up now. The power distribution system is now dual and the electronic controls are on backed-up parallel processing computers.

"One other matter needs to be discussed. We have a miniature nuclear power plant on Mars. It is controled by computers and automated equipment. That system has been working and will continue to work until the colonists arrive. In the event of an urgent emergency the system is designed to withdraw the fuel rods and drop them into a deep builtin well which can be automatically flooded with water while repairs are made. The reactor chamber is reinforced and would be flooded with subfreezing Mars atmosphere if needed.

"Robots will make repairs to the machinery in the chamber and will take other steps as needed. This is a small and easy to maintain design. If needed the radioactive material can be removed to the original outside site dug into the ground.

The president turned to his science advisor and said, "I hope someone understood what he just said."

The advisor smiled.

"Okay, you tell me. What do you consider the weak point?"

Rob said, "The structural integrity of the passageways and the domed rooms can't be tested except at Meteor Crater and in simulations. All of the material we are cutting through has been frozen solid for billions of years. We will be heating that space, which will expand the material adjacent to it. That should just make the fit tighter and more solid, but we just do not have experience. Remember that all of our construction involves Marscrete stone block arches and domes that also atmospherically seal the interior.

"We have a passageway extending north and south along the Outer Ring and another inward along the West Radial. Right now, those are dead ends and a tunnel collapse could trap anything. Our first construction effort this next year is to extend the Outer Ring passageways toward the Northwest and Southwest Radials, but we will build emergency tunnel exits when we reach the middle. We will also extend the West Radial inward past the Inner Ring and create an emergency exit leading to the depression in the middle of the impact crater. Still there are some risks we just cannot avoid."

"Are the ships ready and what actions would you take over these next three months to prepare for launch?"

"Dr. Randall will discuss the preparedness of the ships."

Tim activated his multimedia presentation. "Mr. President, we have six ships available—of which four would actually go to Mars. Those four are thoroughly tested and are now substantially loaded for launch. The other two are spares. They are nearly identical to those that were launched twenty-one months ago. Some equipment has been modified to resolve some problems we encountered during their transit and landing. The *Ares* ships were our test and generally performed well."

The president looked at Mary. He said, "Your plants and animals took a beating. What have you learned?"

"Sir, we had three major problems. The *Fusion* crash-landed and many animals suffered injury from either the impact or decompression. We certainly want to avoid that experience and have taken precautions for our own journey." That reference drove the point home to those present that they were actually speaking with the colonists. They shifted a bit in their seats.

"The second problem was the extended stay of the plants and animals in the confines of the landers. We will not depart if we do not have space ready to receive us on Mars.

"The third problem was keeping the chicken eggs virtually frozen and not incubated for so long that very few hatched. The eggs we are taking with us will be re-enclosed inside artificial shells that can tolerate freezing and thawing. Samples have been cycled with hatching under some rather bizarre conditions. We should fare much better this time.

"We are also taking some new animals. We are taking some young adult female sheep that we will implant with frozen fertilized eggs once they are on Mars. Young animals do not mature naturally in zero gravity so they cannot be taken. Sheep can provide milk, wool, and even meat. We are also taking pigs."

The interrogation continued for two and a half hours with everyone responding to very pointed questions. Then the president shifted the discussion into positive interests. He spoke directly to the team of four.

"Are all of you prepared for this experience? I hope you get along together."

Joan said, "Quite frankly, we have become so immersed in our work that we've had little time to reflect. We've become more like family than colleagues. Still, it will be different when we are entirely alone with each other. We are married couples—married to each other and to our mission. We certainly know our own foibles and our friends' idiosyncrasies and we accept them as personality traits—even love them because of those behaviors.

"The mission has become our lives. We might as well have been on Mars the past fifteen months, what with all the time we have spent directing Mars-located activities and examining the results. The major difference will be confinement inside of the Craterwall Base most of the time and not within the JPL control center. We will not have the freedom to go outside and see blue skies. The advantage will be that we can directly inspect our work when needed. We will be working in real time on outside activities conducted by robots and AI equipment. There will be no radio transmission delays except for the continuing constant support from JPL.

"We will miss our friends here, but we will not be out of contact. We will have pseudo-verbal conversations in delayed transmission. We will also look forward to the next fleet of friends who will join us in two more years. We will always have all of you to provide us with encouragement."

That cinched it. The mission would be a go if they could only do what they had promised.

Chapter 15:

Interplanetary Journey /
Pervasive Dust Storm

THE SHIPS THAT would carry the first humans to another planet were *Venture 1, Venture 2, Venture 3,* and *Venture 4.* However, the ships would not travel together since they had to land at safe intervals and there must be time between launches to ensure that each ship was thoroughly inspected. *Venture 1* and *Venture 2* would each carry a man and a woman. The other two ships would carry equipment, supplies, animals, and plants. *Venture 1* would also carry two robots to assist the humans in reaching Craterwall and carry critical repair parts and supplies. Craterwall would be prepared for their arrival with basic self-sufficiency facilities and processes as well as a nuclear power system, a life-support system, and animal and plant life that would have survived their first Mars fall-winter, the frigid season in the northern hemisphere.

Everyone assembled at Cape Canaveral well in advance of the launch date. The cargo pods were fully assembled and attached beneath each lander. Each ship contained a command module where the crew and all living things would live in a life-support environment during the entire journey.

The four team members stood together on the launch pad away from the bustle of the area. They looked up at the giant *Ares* rocket; the *Venture 1* crew module seemed so remote on its tip. The gantry surrounding the ship only made it seem more surreal.

Tim took Joan's hand and said, "It's magnificent, isn't it? Looking at this ship makes me feel small—insignificant compared to the universe."

Joan squeezed his hand. "I might feel frightened if it were not for you going with me. You make me feel confident that we can overcome whatever may await us."

Rob and Mary could see their own ship on another launch pad. Rob squeezed Mary's shoulders. Mary had her arm around his waist. "The next time we're here, we'll be in our astronaut suits. This is really so huge—it dwarfs us."

Both couples had spent time with family and friends before moving to the launch area and many of those loved ones had decided to join them for the launch.

*   *   *

Gus was giving the press an update before the launch.

"The weather on Mars will begin to warm up as the planet approaches its northern hemisphere spring. Five months ago, the planet was entering its winter season so we stopped all outside activity, including all excavation within the crater wall. It was necessary since excavation required transporting the excavated spoil and dumping it outside the crater where the temperature is now as cold as our South Pole.

"The surface will become warm enough to resume some excavation activity about three months from now—as our first two *Venture* ships reach the midpoint of their journeys. A segment of the Outer Ring now connects the ten enclosed and heated electricity-generation and materials-processing facilities. That means we have constructed ten domed rooms that are each eight meters across. We have also constructed 160 meters of the arched walkway. That space is pressurized with Mars atmosphere compressed to 80 percent of Earth's sea level atmospheric pressure and heated. Only the single habitat room has been maintained with an oxygen, nitrogen, and carbon dioxide atmosphere.

"Again, we ask that you keep our activities in perspective. This is all extremely dangerous work and it is very new for all of us. What would be simple here on Earth will be far from simple on Mars. We send our four astronauts to Mars with our best wishes and earnest desire for their safe journey and success in all of their undertakings. We know that we speak for everyone in that regard."

*   *   *

Everything that would land on Mars had been sanitized and placed in isolation for two weeks before the launch. Tim and Joan seemed lonely as they walked to the ships in their suits and rode up in the gantry elevators.

The pre-launch countdown had stopped while the first couple was settled into their seats and strapped down for the launch. The portals were closed and sealed. They were alone in the small space where they would live for the next six months. They were Adam and Eve going across the solar system to settle humans on a second world. There was no more lonely experience. The burden of their commitment finally rested fully on their shoulders. They looked at each other and touched gloved hands. The countdown resumed. The last few digits were recited in typical NASA countdown fashion.

Instruments were measuring all of their critical equipment functions for the split second when computers would decide that everything was ready for the journey. The huge booster rockets ignited and their journey began. The glare of the rockets brilliantly lit the entire scene. The acceleration was slight at the outset, but increased to nearly four G's before the rockets separated.

Their interplanetary-propulsion rockets ignited to increase their velocity to over 40,000 miles per hour. They took a course near the Moon and then assumed a path that would become the unique elliptical solar orbit that was tangent to the Earth-moon orbit and ultimately the Mars orbit. The die was cast. They were committed.

Tim and Joan recited a list of affirming reports that they were okay. Since the ship was designed with multiple backup systems, everyone in launch control was checking on all systems throughout the ship after the pressures of the launch.

They were being sent with the rudiments for survival, but there had been special gifts. A carton of finch eggs would hatch into small colorful birds that could flit about in the vegetative portion of the Craterwall habitat. Finches were selected because Charles Darwin did much of his study that led to his theory of evolution using finches. A male and a female ferret would share the cabin with them during the journey. They were placed in a hamster-type race that would spin during the entire journey to provide a mite of gravitational orientation—without which the creatures could well suffer from zero-gravity disorientation. The ferrets had been carried in special pouches attached by Velcro to the front of the human spacesuits during the rocket firings.

Their first checks were of the animals they carried. Two adult female pigs were secured in harnesses during the launch and squealed unmercifully after their increased weight during launch went to nothing and they found themselves floating about. Joan transferred them to a centrifugal cage that gave them just enough spin to calm them down and keep their waste in the collection pan. Fingerling fish were spun in their tanks as were shrimp and oysters.

Joan laughed. "Tim, you need to look at this. Our critters are adjusting to their new environment. Kinda makes your head spin, doesn't it?"

He laughed in response. "Can you reach the video camera? This is too good to keep to ourselves. Mary and Rob should know what to expect!"

Mary and Rob would be carrying trout, bass, hives of unhatched honeybees, and two nanny goats. *Venture 3* and *Venture 4* carried fish and bird eggs as well as less critical mammals.

As they went past the Moon, they were treated to the first close, personal view since the lunar landings almost a half century earlier. Looking back toward Earth reminded them of the Christmas *Apollo 10* flyby and the comment then that Earth from this distance looked much like a Christmas tree ornament. This maneuver also bent their trajectory to send then onward to Mars, but only after six months in space. Mirrors and television cameras onboard gave them a full three-dimensional view of the star fields of the heavens. Mars looked like a bright star-like rosy pinprick of light to the naked eye. Tim said a silent prayer that their course was calculated and would execute correctly for their long journey.

The Moon faded behind them—along with their vision of Earth. They had already gone farther from Earth than any humans had gone. They were bidding farewell to not just a place but the foundation of all their hopes and dreams. A sinking feeling accompanied their final farewell.

As the hours sped past, they saw their familiar place in the heavens grow smaller and fainter in the distance. Joan turned to her partner and reached out for him. It was the biggest gamble by any humans since the dawn of time. The sun was fiercely bright, and Earth and the Moon were becoming ever smaller disks against an ebony background. Never had any humans been so alone since Homo sapiens walked the planet. They felt the need to comfort each other as they sped ever farther from their native land. It was not just the obvious danger; it was the solitude that no one had known before.

There was little to be heard from Earth Control—the new call sign for their comrades in this greatest of all adventures. Somehow, the electronic media sensed their situation more and more as the time of transmission from *Venture 1* to Earth increased to more and more seconds, then minutes. There was little to tell the waiting billions on Earth as they coasted onward ... just another twenty-four hours, food from sealed packages injected with precious water, animals that were becoming more their companions each day, a routine of exercise, checking instruments, and confirming that all was well.

Joan and Tim sent a very special message wishing Mary and Rob well in their launch. The long time in isolation may continue, but they would have companions in space.

The launch of *Venture 2* had one hold delay while an electronics system was repaired—and then Mary and Rob were thrust outward from the womb and turmoil of humanity to follow their friends in pursuit of human destiny.

From that point the two crews communicated with each other through Earth Control. It was a party line of sorts that allowed everyone to share in the isolation of space flight and the companionship of those four who shared the great quest for a new human home.

Not long into their journey, Joan and Tim explored the most personal experience of all. Intimacy in space was not new, but in total privacy and zero gravity, there was opportunity to explore more and more aspects. They were required to exercise extensively to keep their bodies in shape. They soon discovered that they could combine their zero gravity exercise with intimacy, which they came to call sexercize. Only the ferrets would ever know of the touch and feel and excitement and complete bonding experienced and expressed to each other by two lonely humans.

The launches for unmanned *Venture 3* and *Venture 4* went off with only minor hitches. The cargo ships were their insurance that the colonists could assemble the remainder of the Craterwall base. They carried the redundancy and the tools needed to convert more native Mars resources into new and repaired equipment that would make the base into a colony.

Months later, as their ship moved farther outward from the sun, its orbital speed would decrease and Earth would move past them in the clockwork of the planets. By then, they would sense an increase in the brilliance and size of their new home, a rose-colored beacon that urged them onward. Mars would become their point of contact with reality.

Venture 1 and *Venture 2* had made a small rocket firing mid-course correction—a tweak to better align them for their landing. Ever so slowly, their speed diminished until they were firmly gripped by the gravitational pull of Mars.

*   *   *

Dust storms and other weather conditions on Mars were monitored by JPL. Three months into their journey, a small spiral of dust appeared at the edge of the North Polar ice cap. As time wore on, the spiral grew and extended southward and a bit westward across the Chryse Acidalia until it neared the Genesis Crater.

Meanwhile, JPL continued the construction work by the robots on Mars. The processing equipment on the Outer Ring was successfully reactivated. Interior construction resumed at the second month with carving the West Radial eastward to the Inner Ring and then a short distance northward around the ring. By the middle of the third month, excavation and Marscrete lining of the first two rooms was underway. Excavation of the next two bio rooms commenced at the end of the fourth month while utilities were being installed along the West Radial and the stub of the Inner Ring. By the middle of the fifth month of the interplanetary journey, the second two bio rooms were completed and construction began on the fifth and sixth bio rooms on the Inner Ring while utilities and habitat atmospheric control equipment were being installed elsewhere.

The construction was all according to plan, but the dust storm was not. Gus sent a text message to what was now being called the Mars Team on their two *Venture* ships.

Mars Team alert: the dust storm being generated from the mars North Pole is extending below 50 degrees north and expanding in the direction of Craterwall. We anticipate that it should reach Craterwall in about ten mars days. You should anticipate loss of radio contact with Craterwall and that your landing will be in extremely low-visibility conditions. You will not be able to communicate with the Mars GPS, Comsat, or *Venture 2* from Craterwall once *Venture 1* is landed.

Also note that we expect taking the Alpha generator intermittently out of operation. We will be reducing the electrical load on the Beta generator to minimums needed to sustain life support and radio communications.

The dust storm enveloped the entire Craterwall region a month before the *Venture 1* landing. When communications with the Mars surface failed, there could be no more direction of the planned construction from JPL. *Venture 1* would be landing in the midst of the swirling dust storm.

All construction operations stopped. The processing equipment had already been deactivated. However, the second generator was acting up as well. It was necessary to provide power for the environmental systems. Bearings on Alpha had been running quite hot before it was turned off and those on Beta were heating up as well. Alpha was turned off intermittently to cool placing the entire electrical load on Beta during these periods.

Mars was located near the inside edge of the asteroid belt. The space around Mars was normally experiencing double the number of meteorites as was experienced near Earth. The number was still quite minute, but there was still the potential of a chance encounter for approaching spacecraft. However, one day before *Venture 1's* planned descent, *Venture 2* was punctured by a very small meteorite. The ship echoed from the hit and then an alarm screamed.

Rob yelled, "What the hell? We just lost our life-support air, Mary! Seal your jumpsuit now!"

The jumpsuits were designed for light-duty pressurization. A hood sealed quickly into place and snap air connections quickly inflated the suits, bringing them up to normal pressure. Alarms spread to Earth and *Venture 1* at the speed of light. Trace dyes automatically followed the flow of air to the hull ruptures.

Rob had grabbed a patch and was locating the first rupture at the entry point while Joan had another patch and was applying her patch to the exit hole. The fiberglass reinforced, self-adhesive patch was slapped into place, and then a second and larger patch was carefully applied over the first.

Pumps sought to retrieve as much of their atmosphere as possible and store it in tanks as the crew members finished putting on their pressure suits. Rob and Mary had suited quickly. Both were already bulbous in their inflated coveralls.

Meanwhile the four astrocolonists conferred much more rapidly across a much shorter distance than with Earth. Rob recited oxygen pressure instrument readings to Tim, gauging just how much life resource remained.

"Rob, you and Mary need to survive at least eleven days to reach our landing zone. The animals and plants have self-sufficient air and food supplies, but you still need to monitor that."

Mary asked, "Do you think we'll make it?"

Tim wasn't sure. He said, "The question is whether the shock of the double impact might have shaken something loose in the hull or among our instruments. There could be some small leaks that would siphon air through the hull." Tim hesitated a little too long. "You have enough oxygen if this is the end of the story. We will continue talking with you until we penetrate the dust cloud. Remember, we can land your ship remotely to set you down in an open space if we can regain radio contact from the surface. You will not hear from us until you penetrate the dust storm and approach the surface. Stay in contact with Gus and Earth Control. If a leak does materialize, they're your experts. Remember that *Apollo 13* made it to safety with a bunch like them."

Rob asked, "So we won't know if you've landed safely until the last minute of our landing?"

"We'll be here for you, Rob. You can count on it."

Rob could hear some doubt in Tim's voice.

*   *   *

The time came for *Venture 1* to begin its approach and landing. Tim sent a special message through Marscom to *Venture 2*.

"Mary and Rob, this is our last message before we begin our descent. We want you to know that we love you both and look forward to joining you on the surface. We will be using geostationary GPS in our approach, but that is still radio and cannot penetrate the dust blanket either. We'll switch to radar when we get close enough to the ground. Once we're on the ground, we'll be out of contact. We'll make sure the radio beacon at the water well is operating to welcome you.

"I'll set up our ship as a remote control to bring you down. We know exactly when you should be arriving and on what course. You should anticipate that we'll take control of your ship and land you near our location. We'll be providing air to your main life-support system through your ship's external connection as soon as we reach you."

Joan added, "We need your companionship. We must sign off now. We wish you Godspeed and will welcome you with open arms."

Rob and Mary were apprehensive about this unproven method of getting them safely down. The *Thor* ships had landed without human guidance—but they would feel much better with Tim helping out.

"Joan and Tim, we wish you a very happy landing. Your contact over these many months has meant everything to us. We love you both. We know that somehow we will make this work."

When radio contact was broken, Joan looked directly into Tim's eyes. "I'm worried—very worried—not so much about us as about Rob and Mary."

Tim clenched his teeth. He said, "Worst case? A lot depends upon both of us being almost exactly on target for the landing oval. I can get us down okay even if we are off target, but I won't be able to help our friends land from the ground if both ships are not near the well. They'll have a chance if the Omni beacon at the well is working. They will still have a chance as good as the *Thor* ships—no matter where they come down. It's a crapshoot."

"You didn't have to be so blunt."

"Remember I said worst case. My money is on us. Once we get to Craterwall, we can work some magic to help Rob and Mary. That is why it's so important to have people on Mars."

Joan said, "I know."

CHAPTER 16:
FIRST PEOPLE ON MARS

VENTURE 1 **WAS** approaching Mars after its six-month journey. All systems were checked in preparation for the landing. The deceleration burn would slow their craft, placing them in position to enter their descent profile for a straight landing. Their final burn would slow the craft and control the vertical descent to the Craterwall landing site. Landings on Mars were always tricky.

Tim said, "Joan, it's about time to become astronauts again. We should put on our suits. I'll help you get into yours."

They had been traveling in light coveralls. Getting into the suit required a moderate amount of dexterity, particularly without gravity to hold things in place. Tim's help was useful.

Joan smiled weakly. "Am I an astronaut again?"

Tim nodded and began assembling his own pressure suit. Their journey had been spent in low atmospheric pressure. That was standard in space flights, the space station, and the new base on Mars. The lower pressure made it possible to move about readily in space or on the surface of Mars without the risk of a "bends" experience. Those could be quite tragic. They resulted from a quick reduction in ambient air pressure, originally experienced when deep sea divers were brought to the surface too quickly. Oxygen and carbon dioxide levels in the lower pressure mix were adjusted so that normal aspiration would breathe in the right amount of oxygen to the lungs.

Tim and Joan sealed their helmets and pressurized their suits with some overpressure to test their garments for airtight function. All was okay. They removed their helmets until the actual descent. The suit was to protect them in the event of a rupture in the ship's hull upon landing and in preparation for their exodus onto the surface of the planet.

They were descending straight ahead, using gyroscopically stabilized instruments, and then flying blind until they got close enough to the ground to use radar to track their position over the surface. All equipment was operational and they were on descent profile. The situation looked good for a safe landing.

Tim spotted the summit of the 25,000-meter Olympus Mons volcano. In the distance, the triplet craters of the Tharsis Montes matched the height of Olympus Mons, but being surrounded by the mass of the bulge, they somehow seemed smaller. The terrain below the ship was clear to view, up to the distant smudge on the horizon at their targeted landing oval, which was obscured by the dust cloud.

Tim reported through the Marscom for relay to Earth Control until they entered the dust cloud that blocked their transmission. He said, "The drag chutes have automatically deployed. Descent is on profile. We're firing our landing rockets and holding our chutes until we turn more vertical. We now have a faint radar image of the surface. We are not picking up the navigational Omni beacon marking the well site. Radar reports we're at three hundred meters and I'm shifting north to avoid the well and nearby power lines. I have visual of the surface, released the chutes, and see the well. We were right on top of it." A billow of surface dust enveloped the craft. "We're over near level terrain and settling, touching down ... all rockets off." There was a pause. "Control, *Venture* has landed."

Joan let out her breath. She said, "Great landing! However, the radio showed you had lost your link to Marscom about the time you released the braking array and deployed the parachutes. I couldn't see the well."

Tim said, "Thanks, but be careful of our new consequential weight from Mars gravity. We spent a long time in space at zero G."

"Will do. The well was on my side of the ship, but I can hardly see it now. There was supposed to be a radio link at the well that doesn't appear to be working. My message was recorded and will be transmitted as soon as a pulse is received from the satellite.

"We need to reach the well to make a ground com connection and find out how things are holding up at Craterwall."

Joan had grown more concerned about the dense dust. They were ten kilometers from the safety of their base, placing them in a broad area that was relatively clear of large boulders. The atmospheric dust obscured vision as heavy fog would.

They had one electronic safety link to the beacon. Tim fired a powerful radio pulse, a ping, from the ship to invite a response from the beacon. The beacon responded with a powerful radio blast providing them with the distance from the beacon and its azimuth to their ship. Their navigation computer used that to orient its electronics to topographic north and shared this information with all equipment onboard *Venture*. They now knew their exact position and orientation.

Tim ordered firing of the exploding bolts holding the cargo pod in place. The pod slowly lowered on cables until reaching the surface. All was going according to plan. Two AI robots unfolded, found purchase within the pod, and then stepped onto the surface. This was like the landings two Earth-years before. Ramps were extended and the first vehicle, a hauler loaded with immediately needed supplies, repair parts, and equipment rolled down onto the surface alongside the robots.

Tim and Joan were executing a many-times-rehearsed procedure intended to lead them to safety. So far, they could pause without loss of critical time. Then pumps sucked the atmosphere from inside the crew cabin into cylinders. Their pressure suits ballooned as the cabin pressure diminished. The hatch unlocked on command and opened when the pressure between the Mars atmosphere and the cabin equalized. Tim tugged the individual life-support shells containing their live cargo into position for discharge to the waiting robots. The pods were easily carried to the waiting hauler and snapped into place on the hauler's bed. They were then connected by hose to the hauler's life-support system. Only then did Tim climb down the ladder to the surface, followed by Joan. They spiked a flag into the surface and made remarks that would be long remembered about the time when men first set foot on this seemingly barren world.

"We come, man and woman as partners, to build a new home for humanity. We pledge to respect this place and its natural order of things. We make this commitment for ourselves, our fellow astrocolonists, and all those who shall follow."

But the clock was running. They climbed into the truck's cab, but they didn't flood it with life-support gases. Tim quickly steered the vehicle over to the well site. The radio still could not make connection to the communications transmitter at the well site. Tim directed a robot to extend a communications cable and connect to an outlet at the base of the nearby electrical power line stanchion. Only then did the fiber optic com link come alive to the interior Craterwall Base. Standard data queries and responses soon informed their receivers as to the status of all equipment at the well site and Craterwall base. Specific data showing malfunctions soon displayed on the inside of their space helmets.

Joan said, "It looks like the power line failed at stanchion 3.48 and power was turned off from base. The well here is at 9.93 kilometers. The shutdown of the well was incomplete, but continued on fuel cell to completion two days ago.

"We knew the generators at the power plant were failing. One generator was turned off two weeks ago to save it from further physical failure of its bearings and perhaps catastrophic damage to the armature wrapped in fine-wound electrical wires. The other generator continued to produce electricity in a limited way … enough to keep the base's life support and grow lighting operational. We've arrived none too soon."

Tim saw a much more dangerous situation. "They cut off all construction two weeks ago! The sixth room was never completed and only the south section of the Outer Ring has MHA, habitat atmosphere. The processing of atmospherics and fuel in the rooms off of the north section was stopped. There is very little atmospheric reserve."

Joan was grasping their position. She said, "They never moved the biologicals from that single room on the Outer Ring. I see the robots are stored in the new room across the passageway. That's the only other livable space we can use."

Tim ordered a backhoe/loader from Craterwall to meet them at the cable break. The robots were designed to appear relatively human—one head, two arms, two legs, shoulders and pelvis—much like the design used in *I, Robot*. They also could communicate electronically or orally in human language. They were even given human names. One of their two new robots, Carol, was left at the well. That one would monitor operations at the well when power was restored and then return to the safety of the ship. Ivan, the second robot, would come with them. The loader then activated life-support for its cab. The loader was to be occupied by Becky, a robot from the *Thor* mission. Tim headed the hauler down the trail.

Joan assessed their situation. She said, "Tim, is this as bad as I think it is?"

Tim shook his head. "We don't have any time to spare. The generators and the atmospherics are critical. We're not inside the Craterwall yet and a lot of the passageways and domed rooms are dangerously incomplete. There are more loose ends than I can count. We'll be lucky to have the situation somewhat stabilized before *Venture 2* arrives. But I don't see anything that we can't handle with a little luck!" He wanted to be positive for Joan.

They had been creeping along the Power Line Trail for the better part of an hour. The vehicles reached the stanchion identified as the location of the electrical failure. Ivan climbed into the bucket of the loader and was lifted to the top of the stanchion. Dust had accumulated between the power cable and the stanchion, causing a short circuit. The robot brushed aside the dust, but the cable was held together by a thread. An electrical cable splice was needed and inserted. Ivan was then lowered to the ground.

Tim ordered the command computer at the base to restart power to the well site. Receipt of power was shortly confirmed by Carol.

She immediately responded, "The well stem is clear of moisture. I suggest waiting until the off-line generator is repaired before turning on more power. All that is powered now is the Omni beacon."

Ten minutes later, the hauler made connection by radio with Craterwall and homed on the Omni radio beacon at the main tunnel entrance. The hauler entered, but the backhoe was parked at the dock outside the air lock. Tim and Joan continued the short distance up the outside passageway to the vehicle dock at the main air lock. The outer door opened upon command and they found the inner door already open. The interior of all the passageways was at Mars standard atmosphere.

The hauler stopped outside the air lock and then slowly eased through. The small hauler just barely fit through the opening.

Tim and Joan knew that Earth Control had been busy during the three months before their arrival. The entire West Radial leading back to the planned living space had been excavated and reinforced with Marscrete blocks, sealed with a paste of Marscrete, and then fired. It was ready for pressure testing, equaling a habitat atmosphere, but it was still at Mars atmosphere.

A stub passageway along the Inner Ring off the West Radial had been constructed and a wing of five habitat rooms excavated, lined with Marscrete blocks, and sealed, but had not yet been tested. A sixth room was enclosed with Marscrete blocks, but utilities for that room were not yet installed—and it was not sealed.

All of this work by the robots had been done during the dust storm without any direction from Earth Control. That was rather remarkable. Still, the construction needed to be carefully inspected for any dangerous oversights before a habitat atmosphere could be added.

Local hours and minutes on Mars were maintained the same as on Earth to avoid confusion. That meant Mars had just over a twenty-four-hour day with the leftover piece after the twenty-fourth hour. Noon was placed at solar midday.

Time was slipping away; it had been almost three hours since landing. They were faring poorly under their newfound weight, but they knew time was critical and struggled on.

Tim and Joan entered the south leading power plant wing of the Outer Ring through the only operational air lock inside Craterwall. The loaded hauler also barely fit through that lock and then they felt the interior pressure matching their suit pressure. They shed their pressure suits and drove the hauler down the corridor. They passed the only habitat room—the one that housed the plants and animals from the *Thor* mission. They continued past the nuclear furnace and its warning of interior radiation.

The operating Beta generator was in the room with the steam engine and the idle Alpha generator, which together were their intended source of power. Beta emitted a loud squealing noise as it struggled to keep operating. They were clearly in a race against time to get the Alpha into operation before the Beta self-destructed.

Ivan and Becky unloaded the bearings and other overhaul parts for the two generators. They were placed on a plastic tarp marked with silhouettes for each that part they would need during the overhaul. The original gin pole was already set up over the inactive Alpha. It would be needed to raise the upper frame of the generator and lift out the Alpha's armature, the rotating axial part that was tightly wound with thin wires and that cut through the magnetic field to cause electrons to be pushed down the wires to create an electrical current.

Joan took the hauler back to the bio habitat room just inside the Outer Ring air lock. She was joined there by two other robots, Elsie and Jeb. That room contained all of the plants and animals from the *Thor* mission. She directed the robots to move all of the inactive robots from the storage room across the Outer Ring. She then had them move all of the animals from the habitat room to the storage room across the corridor containing their equipment and supplies.

Joan had Elsie and Jeb unload the new plants from the hauler into the vacated space in the original habitat room, which had grow lights and water. The livestock was unloaded in the new livestock room. All of the plants were in grow light and watered space including hydroponics. The animals were in a separate improvised living space across the corridor.

Tim had directed Ivan and Becky to unbolt the top of the cool Alpha generator. They used the gin pole to remove the top and then raised the armature high enough to clear the frame. Everything was done in slow motion. Ivan and Becky pulled the bearings off of the armature's shaft with an extractor tool. Each bearing seat required more than an hour to smooth with an emery cloth and then the robots slid the new bearings up the shaft and secured them in position.

The refurbished armature was then lowered into the generator frame, tested for free movement, and the top was lowered into position and firmly bolted down. The armature and bearing repair required more than three hours of work.

The original electrical harnesses were replaced as was the generator's computer control with its electrical distribution system. Finally, Tim linked the generator to the steam turbine and spun it up without electrical load. In just under five hours, the generator was repaired and reassembled.

Joan walked into the generator room to see Tim inspecting his completed work. She placed her arms around his shoulders and kissed him on the cheek. The squeal of the Beta generator seemed to be even louder than before, pleading to be relieved of its burden.

Tim and Joan stepped out of the room, smiling. Joan placed her hands over her ears and yelled, "Can't you turn off the other generator?"

"Just a little while longer. I want to be sure the Alpha is working reliably."

"If the Beta blows, we could lose our backup."

"You have a point."

The Alpha generator was gradually loaded with increased electrical power usage and run for fifteen minutes until it had assumed the full electrical load. Tim stood by for another fifteen minutes of full-load power generation by Alpha.

Beta was taken offline then allowed to spin down and become silent. It would be allowed to cool overnight before repair. The sturdy purr of the Alpha was music to Tim's ears as he wearily walked the distance to the two habitat rooms just beyond the nuclear furnace.

Joan had improvised walled animal enclosures and set out food and water. The newly arrived animals were intimidated by the new pull of gravity. Joan was nudging them to approach the food and water to get them on their feet. Still they were reluctant. The animals from *Thor* were active though. Their example was necessary.

Tim and Joan walked over to where Joan had set up some boxes for seats. They sat with their backs to the stone walls, too tired to care. Still, they held hands and were content. It was evening, although there was no outside light to show it. Tim spotted two air mattresses and inflated them on the floor. Beyond exhaustion, they fell upon a mattress in a stupor. They didn't fall asleep for hours; they wavered in and out of consciousness until they eventually fell into a fitful sleep.

CHAPTER 17:

AN ALIEN WORLD

JOAN NUDGED TIM awake after only five hours of sleep. They were both so sore from the first day's activity that they could hardly move. The next thing Tim knew, he was smelling coffee drifting beneath his nose. Then he sensed the pungent odor one might find in a zoo. His eyes opened just enough to see a steaming cup and a smiling face. He rolled over and sat up slowly.

Joan handed him the cup. She said, "Good morning. We have a busy day ahead. Thank you for fixing the generator yesterday. You saved our lives."

Tim gave her a peck on the cheek. "Don't remind me. Yes, we have an awful lot more to do. Oh! That hurts."

"Let me help you get started. Thankfully we have a microwave oven. Here's a bowl of hot oatmeal—something that's not MRE for the first time in six months. We'll do better for lunch. At least we have a water tap and a drain. I'm afraid our plan was for toilet facilities on the Inner Ring. Just join the animals for now. The robots clean up."

"Ugh. Not even a basic latrine. We do have a long way to go."

Tim was not putting on when he pulled himself to his feet. He was a little dizzy. It seemed that the hours of exercise on the ship had done little good.

Joan was only slightly better, but she was concerned. "It's your heart. It needs to adjust to pumping blood against gravity. Please take it easy."

"Need to get going. Beta awaits me." He switched on his radio link to the robots and put a cell phone I/O device in his ear.

"Ivan and Becky, get everything ready to overhaul the Beta."

The response was immediate. "We have that ready for you. We've moved the gin pole and removed Beta's cover. We can lift out the armature as soon as you arrive."

There was a pair of one-person Segway® transporters at the entrance. Tim climbed aboard and leaned into his desired direction of motion. The Segway moved forward. Tim had learned its operation as part of his training. He let the machine scoot him down the corridor to Beta. He stepped off and simply gave an order for the robots to commence disassembly following the previous day's procedure. He leaned on a spectator chair that raised him above the action.

There was no learning curve on how to repair the generator. The AI systems were remarkably adaptable. The previous day had raised their confidence. Ivan and Becky moved from step to step with Tim just acknowledging the completions. Three hours later, Beta was again producing electricity. Tim directed the Alpha to ease back until Beta became an independent power supply.

Meanwhile, Joan was checking on the reserve status of life-support gases and fuel supplies. They had depleted during the period when the generation systems had been shut down. She fixed a light lunch and carried it to Tim. She found him moving about the power plant facilities. Replacement components, sensors, and remote-controlled valves and pumps were scattered about the rooms waiting to be installed. All of the components were redundant. Thomas and Albert—or Tom and Al—had joined Ivan and Becky. They worked together, moving around the room while Tim followed their activities.

Joan said, "I've checked our supplies. We're very low. Nitrogen is extremely low. We should consider adjusting our task priorities."

Tim knew that the nitrogen was only 3 percent of the Mars atmosphere. The nitrogen would remain a gas after they froze the carbon dioxide, but it could become a liquid upon further refrigeration.

Tim reluctantly agreed to mild physical exertion. "Okay. We seem to be in pretty good shape with the steam lines for now. Let's make an on-site inspection of Craterwall."

They mounted their Segways, stopped at their small piece of the animal habitat to put on their pressure suits, and entered the airlock. Ivan and Becky stood beside them. The suits ballooned. They experienced the very light outside atmospheric pressure when the outer doors opened to the radial corridor. The humans leaned to cross the West Radial, through the open air lock on the other side and into the northwest section of the Outer Ring.

In the Processing Wing, wall lights were triggered by motion sensors and showed the way through the tunnel. The first room on the left was the Marscrete mixing and stone block fusing room. Tim made a mental note that this Marscrete would seal the remainder of the Habitat Wing and hoped that there was enough. A short inspection showed everything in place. They moved on to the most distant of the five additional rooms, the water distillation unit.

Tim was satisfied. He said, "Joan, just go on with Becky and check out the Habitat Wing of the Inner Ring. Give me an item-by-item of what you find. Please don't take any chances. Nothing is critical there at the moment."

Joan and Becky backtracked to the West Radial Corridor and inward along the corridor to the open air lock leading to the unfinished Habitat Wing. Joan had expected to find matters in disarray and they were. The whole area needed to be straightened up before anything could be done. She called for Al and Tom to join them and sort things out in the five finished rooms. The bare rooms were not set up for use.

Joan's greatest concern was the unfinished sixth room. It was completely lined with Marscrete blocks from floor to dome, but there were obvious gaps between the blocks. She called for Nat, Oxy, and Don to join them. She placed Nat, the construction robot, in charge of his fellow robots, filling in the obvious cracks with Marscrete powder and then plastering the walls with Marscrete paste.

Earth Control had directed the continuous work by the robots until the dust storm. Teams of a dozen controllers had rotated between shifts. With no communication from Earth, the guidance of the robots could only come from the humans on Mars. The people could not work twenty-four hour days. They must have rest.

Joan worked alongside her robots and soon discovered that the robots were doing just fine on their own. The robots could continue overnight without direct supervision.

Tim and Ivan repaired the distillation unit and made a thorough inspection before he placed it in operation. Fresh water began accumulating in the distillation tank. Tim left Ivan to monitor the processing and scooted over to the habitat. Joan turned things over to her robots, toured the wing with Tim, and then they headed home to the Power Wing.

Meanwhile, Elsie had tended the vegetation and joined Jeb in nurturing the animals. The new animals still required encouragement to move and eat their food.

The last thing Tim did before retiring was to contact Shirley in *Venture 1* with instructions to begin activating the well first thing in the morning. He would then send Tom and Don out along the Power Line Road to install three new Omni units to gain continuous precision navigation for vehicles traveling the primitive roadway.

Tim and Joan were exhausted, but that had become their normal state. They collapsed to eat a rudimentary meal. They had no energy or time to prepare real food. They whispered their personal thoughts until they fell into a deep sleep in each other's arms.

*   *   *

The third day began with activity in every quarter of the colonial operation. Tom and Don had completed their installation of the Omni beacons by noon and joined Shirley at the wellhead. She had the well in full operation and cloudy water was flowing into the simple freezer unit. The water flowed through a manifold that filled open trays. The water quickly froze in the open air. The trays periodically rotated, dumping the ice into a large box that could be loaded onto a hauler.

The three robots were riding in the *Venture 1* hauler. They loaded the second cargo unit from the lander onto the hauler by hand. The cargo was mainly repair parts, tools for making repairs, new replacement machinery, and parts needed to complete the utility and environmental control systems. Repairing the many pieces of overused machinery was essential. All of the vehicles and robots from the *Thor* mission were long overdue for close maintenance inspection.

It was near day's end when the three robots boarded the hauler and returned to Craterwall. The dust storm was still intense, but their AI hauler easily navigated the chain of new Omni beacons.

Meanwhile, Tim monitored outfitting two rooms in the Habitat Wing as temporary repair shops. One room was for mechanical equipment and the other for robots and electronics. The new cargo from *Venture 1* arrived in the late afternoon. Most of that cargo was placed in the two rooms. The hauler and its cargo were covered with a thick layer of dust, causing Tim to consider constructing a vehicle wash just within the entrance to keep outside dust from spreading within the corridors and rooms.

The joints between the blocks in the sixth room had all been filled tightly. That morning, the surface of the entire room had been covered with a layer of Marscrete paste and the new wall covering was then fired.

Meanwhile Joan saw to resolving the loose ends in the Habitat Wing and inspecting the finished walls, ceilings, and floors.

Tim and Joan finally retired to their space in the temporary animal room in the Power Wing. Amazingly, after just three days in Mars gravity, they were adapted to the half walk-half shuffle within Craterwall. Since they weighed only three-eighths of their Earth weight, they had less traction on the floor. This was compensated by almost sliding across the smooth Marscrete floor so they were less likely to fall.

Eight video monitors scanned nearby outside objects for visibility, testing objects in the landscape. They showed no let up of the continuing storm. They were still very sore from their exertions, but their movements were becoming self assured and more rapid. Their meal that evening included vegetables and rabbit stew. They prepared the meal together. They assembled a folding table just brought in on the hauler. It felt good to at least sit like a human being.

Tim finally smiled after three days of unending exertion. He said, "Mary and Rob will be here in seven more days. We have only four days to get it all working and then two more to set things up."

Joan responded with a smile of her own. "We could make it if nothing screws up. Anyway, I was thinking about how Mary and Rob must be feeling right now. They have no idea whether we landed okay or how they'll get down."

Tim said, "Who knows if those patches in the hull of their ship worked or if any critical equipment was damaged by the meteorite impacts. Computers can test only so much. They were facing a lot of uncertainty."

Joan said, "And that's why we need to do all we can to give them a hearty welcome."

Chapter 18:

Time Runs Short

THEY WERE RUNNING out of time. Rob and Mary would arrive in less than five days.

Tim and Joan worked in the bitter cold and near-vacuum of the Mars atmosphere. They took brief respites in the heated, habitat environment of the power plant and inhabited spaces. Their pressure suits included heaters. In effect, they were performing EVA in a gravity environment. The robots worked in the cold.

Tim guided starting up the processing in the room where the dry ice and the liquid nitrogen were evaporated to gas. It was a relatively simple process, but the fittings had to be tested and some replaced. The gas ended up in high-pressure cylinders made of fiberglass-reinforced epoxy.

By end of the day, Tim crashed into this bed for a long sleep. Joan had many tasks in the Habitat Wing setting up facilities and working on the leakage problem in the sixth room. She then collapsed into her bed as well.

Tim tested and repaired the methane-production equipment that separated distilled water into hydrogen and oxygen. The hydrogen was combined with the carbon dioxide to form the methane and the oxygen was stored in high-pressure bottles. Finally, the equipment was placed into operation and began producing an encouraging increase in their limited methane fuel stock. The new supply of oxygen was being combined with the nitrogen to form enough of the habitat atmosphere to fill the corridors and rooms.

The greatest improvement was in the repair status of the vehicles and the robots. Joints, sensors, and cameras were virtually all replaced on the eight D-model robots and on the single C-model. The C-Model received so many repair parts belonging to D-Models—including its computer processor—that it effectively became another D.

By day's end, all of the interior processing operations were working and being monitored by assigned robots.

Joan made a punch list of tasks to be performed to get the facility in working status. They again made a Mars atmosphere pressure test of the Habitat Wing, finally with a successful result.

It had been another very long day.

*   *   *

Getting an oxygen atmosphere inside of Craterwall and heating the entire place was a challenge. All of the West Radial and the Processing Wing of the Outer Ring now had an oxygen-nitrogen atmosphere of breathable air that would be heated. The air locks to the Power Wing were opened.

The Habitat Wing remained under test with an oxygen atmosphere and retained its air lock. The hydroponics tanks and the trays for soil-supported life were already in place there, but irrigation needed to be added. Pens were assembled with feed and water troughs and drains to clean up after the animals.

All of that wing's distribution and collection systems were completed, including animal waste and white water sewage and potable water, air composition monitoring and automatic adjustments, data and information processing and maintenance, and much more.

No animals or plants would be moved until they were certain that everything was operating properly. The outside mining, road upgrade, ore refining, aluminum smelting, and related chemical formulation would not be placed into operation.

The atmospheric regulation would control the ability to add oxygen or nitrogen to spot locations, but all gases were returned to a central monitoring facility in the livestock room still used by the people. The robot, Oxy, monitored this process. The system functioned and circulated much as an air conditioning system worked on Earth except both mix and temperatures were involved. The question was whether the regulation system would work smoothly.

The process to flush the carbon dioxide atmosphere from the Processing Wing and the West Radial had been accomplished overnight. First thing that morning, the habitat atmosphere was introduced from the tanks of mixed oxygen and nitrogen. The tanks were emptied and the mixing unit began directly introducing freshly mixed atmospheric gas. That continued throughout the day.

What resulted were corridors and rooms that were still quite cold. The first step in heating was to pump hot oxygen-nitrogen atmosphere throughout the plumbing. They then slowly introduced low-pressure steam into the warmed plumbing, using the spent steam from the steam turbine in the generator room. The sensitive process could easily result in a disastrous freezing of the newly introduced steam.

The interior was still warming and the pressure increasing as they retired to the quiet of their sleeping space.

*   *   *

Venture 2 was also having problems, but those resulted from the meteorite strike.

Rob was concerned. He said, "Mary, the oxygen is disappearing at an alarming rate. We need to take more stringent measures. It looks like we have a leak in our reservoir."

Mars was growing in their view, but they didn't have enough oxygen to survive until the landing. They were already rationing themselves, but less consumption was needed.

Earth Control ran through their options with Rob and he said, "It looks like we still must reduce our oxygen. Can we reduce what the animals need to a level where they are barely alive? I really don't want to lose them."

Control said, "You will likely need to do the same. We'll monitor you, but you're right."

The onboard AI computer performed its own analysis. Small bottles of oxygen were associated with rescue equipment. Those could be given directly to the people now. Animal consumption of oxygen was minimized. If they gave the animals more, they would all be at risk.

The AI computers onboard *Venture 2* monitored their health, including their blood-oxygen levels. Their oxygen levels would be adjusted to keep the people barely conscious. They composed an automatic message to Tim and Joan, telling them of their desperate situation. It was high-risk, but it could work—theoretically.

*   *   *

Gus and the Genesis Team were at tether's end. Earth Control was desperate to know what was happening on Mars. Craterwall had been a mess when the dust storm interrupted all contact with the surface. It had been over a month since losing contact with the Craterwall, nine days without contact from the *Venture 1* team as they entered the huge dust storm, and now *Venture*

2 was gone while nearly out of life-support supplies, a crew that may never recover consciousness, and facing a high-risk landing. Earth Control kept sending messages to the Marscom only to get the same reply: Message received by the satellite, but no response from the surface.

Gus and Dave had grown quite close to their astro-colonists. More silent prayers were sent to the great master than just from Gus. Everyone felt the same way. The team on Earth—and most of the population of their native planet—were deeply attached to the astro-colonists and their adventure.

It was Earth Control's first encounter with a large dust storm communications failure—at least during this mission. They had continuous communications for the previous two years while they were using robotics to build the elementary base. There were many things that could go wrong now.

The world media was going crazy with wild stories. There were images of crashed ships and of people lost in the middle of the storm with no way to reach the Craterwall base. Gus could only keep telling the reporters that the dust storm persisted.

When *Venture 1* landed, they had heard Tim's verbal landing sequence down to eight hundred meters altitude. They noticed a very narrow window as *Venture 2* would be landing when they could receive a relayed message from Mars—only if Tim realized this and set up such a message. He needed to arrange an automatic message that would trigger as soon as *Venture 1* contacted *Venture 2*. So Gus had his people set up a relay from *Venture 2* of any messages from the surface received during the *Venture 2* landing. It was improbable, but it might happen.

*   *   *

It was time for Tim and Joan to move the animals and plants to the farm in their new Habitat Wing space. Air locks would be automatically activated if anything adverse developed.

With atmosphere in all sections of Craterwall, it was not necessary to close the air locks. However the lighting was activated only when people or robots were using the corridors and /or rooms. The Habitat Wing was lighted at all times.

The Marscrete stone walls were bare—even stark. The first plants were loaded onto the hauler. The walls echoed as the hauler's wheels rolled into the West Radial, past the Middle Ring intersection, and onto the turn into the Inner Ring.

All of the robots were involved with the move. Joan monitored the hydroponics. Tim monitored the rooted plants. Each plant type went to

95

its own room. Samples were moved first to confirm the viability of the environment. They wanted no surprises.

Algae had been introduced as soon as the environmental stability was confirmed. It would consume the carbon dioxide from the hauler exhaust. The vegetation was moved first from the Outer Ring to the Inner Ring. Hydroponics was tricky to move because of all the interconnected equipment. The rooted plants were placed in the second room. Some larger bushes were independently potted and scattered in gaps between the equipment.

The menagerie was moved in sections. All mammals went into one room. The two pigs were first to make the Noah's Ark transfer, followed by the rabbits. The second room contained tilapia in a slowly moving water trough. The chickens from the *Thor* mission were provided with nesting material that they immediately used to lay eggs. Trays of eggs from Earth had been in the incubator since *Venture 1* had landed. They required about three weeks to hatch, and then they would be kept in the room with the fish.

Joan was barely out of her hauler when farmhand robots began unloading the animals. They knew what to expect. Many of the animals were still lying down, avoiding the pull of gravity. Joan released the ferrets that had been their constant companions when they were in the old animal room.

The other two rooms were for the humans. Joan had moved directly to the living and workspace. One room had a small kitchenette with sink, electric range top, electric broiler, microwave oven, refrigerator/freezer, pantry, and a large storage freezer. There was a table for four with foldaway chairs.

The other half of the living and working space contained counters and electrical outlets for power and communications—computers, displays, and all of the latest working tools. Most of the equipment on the ship could be detached and much of it was stacked on a counter waiting for a position in the work area.

The second human room was for sleeping, bathing, and relaxation. The sleeping space was partitioned for privacy.

Joan took charge of their personal gear and space. Lightweight tables and chairs were organized in the kitchen and work area. They would sleep on tubular frames with taut wire bases with spring anchors to hold the air mattresses.

Suddenly, the impact of their situation closed in upon Joan. She looked around. Then seeking comfort, she walked slowly over to the rabbits. The furry creatures didn't seem to know what to make of Joan. They were already reproducing with about two dozen in the warren; all had adapted to Mars gravity. Edible vegetation littered the floor of their enclosure.

She picked up a grown rabbit and held it to her breast. It was soft and trusting. She held it as she walked back into the animal enclosures. A robot saw her with the rabbit and approached, seeking to return the rabbit to its proper place. She waved it off. She needed the comfort of holding another living creature.

When Tim entered the room, he looked directly into her eyes.

She whispered, "We're here. We are really here inside Craterwall and breathing local air. We made it!"

Tim took her hands. "Welcome to our new home."

Joan shed a tear. "It's a fixer-upper ... but we're safe. We actually made it. I love you so much. You got us here in one piece. So few believed we could really do it. And now we're safe—alone in a strange place on a strange world in the middle of a storm. I'm so relieved—you will never know."

"Yes, this is the first time we can stand up, feel our own weight, and have real living space, not just a capsule. This is real, not just some story. All our work and our long trip have paid off. I love you, too. Far more than you'll ever know."

CHAPTER 19:

TO BE OR NOT

ROB AND MARY were due to land—it was the day of the hoped-for reunion. Tim and Joan had worked so hard and had been too busy to consider that this reunion could be anything other than happy. Then it came down upon them. Ten days before they had parted from their friends and descended into the giant cloud of dust. There had been absolutely no communication with them or with Earth Control since their descent.

Tim had been thinking about any way to communicate with Earth Control. That was when he considered the possibility that *Venture 2* may still have contact with Marscom as it sank into the dust cloud. Joan and Tim composed a very concise message that they would send automatically to *Venture 2* as soon as *Venture 1* had contact.

Rob and Mary had repaired the holes in *Venture 2*'s hull. The ship was controlled by JPL on Earth so it would descend to Mars on schedule, regardless of circumstances inside. The question was whether the automatic systems could find a safe landing location and whether Rob and Mary would still be alive after so much oxygen deprivation.

Tim and Joan climbed aboard their hauler, which contained only emergency supplies. A second hauler carried the gin pole to lift heavy weights in an emergency rescue. They exited Craterwall into the still swirling dust cloud. The haulers made their way down the chain of Omni beacons to the well. Numerous trips had been made this way by robots picking up ice and delivering the ice to the distillery.

Tim stopped at the well to confirm that the beacon was transmitting at maximum power. They left the second hauler at the beacon. Tim turned to the beacon's outbound radial direction that would take them to the silently waiting *Venture 1*. The distance was short. Joan stayed with their hauler and Tim entered the ship to set up the electronic links that would seek the approaching *Venture 2*.

The *Venture* ships could guide each other in an emergency. Tim intended to guide *Venture 2*'s landing as soon as the link was established.

Venture 1 had been dormant since they landed. Tim energized the ship. All of his instruments came alive. He powered the radio that should link to *Venture 2* and performed all of the analytical checks to be sure everything was working. It checked out perfectly.

Tim was still wearing his pressure suit. He called Joan. "I'm okay. So now we wait. Meanwhile, let's go through our emergency procedures. When they come within range, we'll get an instant download of all their critical information. If they download but land elsewhere, we should get enough information to determine their probable landing site. We have two robots on each hauler. Dispatch the other hauler as soon as we get a landing fix. Maintaining air pressure is our first priority. Each hauler has emergency pressure suit patches. In worst case, we push them into the emergency bubble and inflate it."

Joan laughed and said, "Tim, just relax. Everything should go fine. If not we're very, very familiar with the emergency procedures."

Time seemed to drag forever then they heard the alert signal. *Venture 2*'s radio signal reached Tim's receiver and was passed along to Joan. Their computers locked together ... and an emergency light lit on Tim's console. Tim took control of the landing. At the same moment, their automatic message went out to *Venture 2*—and ideally to Earth Control. Meanwhile, a recorded message from *Venture 2* was being received on Mars.

"We are due to land tomorrow morning, but we are extremely short of oxygen. We discovered a small leak in our supply. We have so little oxygen that we have adjusted our use to last until landing by using our small emergency bottles. That means we could be unconscious when you get to us. We're counting on you."

Tim immediately discovered that the ship was coming in high and to the east. He made adjustments to the landing profile to make their landing as close as possible. He lost contact before the actual landing. "Joan, they came down about eight kilometers to the north-northeast. Let me take the lead in the other hauler." The hauler from the beacon site reached Tim and he immediately climbed aboard and headed out. Joan was right behind, moving to intercept the beacon's radial toward the landing site. The terrain was littered with occasional boulders, limiting their speed. Repeatedly Joan threw on her brakes as Tim encountered boulders. The beacon's signal was blocked by terrain and gave out at just five kilometers radial distance.

Tim began dead reckoning using his gyroscopic compass. "I'm not receiving a signal from *Venture 2*."

He swerved east around a large boulder and then swerved north again. He was maneuvering so much that it was hard to keep track of their location. He dropped into a gully , followed that for a few hundred meters, then he climbed out to be in a better position to receive a signal.

Joan signaled. "My compass tumbled. Hold up until I take a setting from you."

Tim slowed to a stop with Joan aligned in parallel. "I read 037 degrees."

Joan responded affirmatively and both started out again.

Tim called, "I have a weak signal northwest." They switched directions into a cloud of dust. The signal disappeared then came back stronger. "Attaboy … stay with me."

The signal held and Tim abruptly stopped. *Venture 2* was right in front of him; it appeared to have landed safely. Joan swung to the left and stopped. Tim said, "Attach your air to the left side and I'll take the right. I'm picking up bio signals. They're barely alive."

The robots needed no instructions. They had been following the conversation and jumped off of the haulers as soon as they were nearly stopped. Ivan snapped the hose to *Venture 2* on his side and Becky did the same on her side. Becky was already forcing oxygen into the external snap connection. They knew all living things were in pressure suits or enclosures. Those were linked by hose connections to their supply, which was now being replenished. Ivan went up the ladder to the cabin hatch. Since there was no pressure showing inside the cabin, he popped the hatch immediately. Becky went up the ladder on the other side and opened that hatch.

The pressure suits had a built-in respiration-assist function that was operating. Critical body functions were monitored at all times; they showed both people with nearly normal mental activity. A medical system was actively treating both people. They were semiconscious. Stimulants were mixed with the air being supplied.

Tim had already fired the cargo bolts on *Venture 2* and lowered their cargo-laden hauler. All of this happened in a few short moments.

Tim and Joan grabbed Rob's and Mary's gloved hands. Tim and Joan were crying.

Tim said, "My God, you gave us a scare. Hang in there. The monitors show you're holding your own."

Joan was leaning in through the other hatch. Rob and Mary were nearly flat in their couches. The bands built into their pressure suits inflated and released, assisting them in breathing.

Joan reached around Mary's pressure suit in a simulated bear hug. "Just stay there for a while. You're feeling gravity for the first time in ages. We don't want to do anything that would make it harder on you. You're safe now and that's all we ask for."

Rob whispered, "How are things here?"

Tim said, "Right now they couldn't be better. All that matters to us is that you are both here and you get back on your feet."

The exertion of the past ten days disappeared. The strain they had endured melted away. Emotions cried out in relief. Rob and Mary were crying, too. Four lonely people stared fondly at each other through plastic domed helmets. It was far from the welcome they had hoped to extend, but it was still a happy reunion.

*　　*　　*

Tim's recorded message had relayed through Marscom and on to Earth Control:

"We're okay. Craterwall is all okay. Generators are overhauled. Habitat completed. Construction completed. Processing systems working. Atmosphere throughout. Send love—"

Then the message was cut off.

Gus had been waiting with the control team to see if they would get any news. He bowed his head and wept. Most of the team members were also choked up.

Finally, Gus said haltingly, "Tim and Joan made it and had some control of *Venture 2*'s landing. Vitals for Mary and Rob would be okay if Tim and Joan reached them in time. We now have people on Mars in a life-supported environment. I thank God."

Everyone was stunned—too relieved to celebrate. No one expected this much good news. Everyone had lived the preparations, the journeys, and their friends' trials.

Gus turned to the press. "Please give us a little while to ourselves. This is too much of a shock, a really good news shock."

*　　*　　*

The robots were not idle at the landing site. Cargo was moved from the ship to the other two haulers. Ivan shut down *Venture 1* and sealed it. Two hours later, the robots tenderly lifted Rob and Mary to the haulers. The robots would drive.

Rob saw the dark dust blocking his view. "We landed in this soup?"

101

Tim laughed in relief. "We're about eighteen kilometers from our home, a good part across rugged country. You're going to experience some bumps. We installed a string of Omni beacons from the well to Craterwall. The robots can backtrack to the well. They have the navigation down cold from that point."

Mary was still in shock. "Earth Control shut down the Craterwall operation when the dust storm interfered. It must have been a mess for you."

Joan turned to her friend. "Look. It's working now and that's what counts."

Rob shook his head. "That would be impossible. Are we really okay?"

Tim smiled. "We and the robots have been very, very busy. It's not perfect. Yesterday we still needed to move all of the plants and animals to the Habitat Wing, but there's air in all of the constructed space. Don't worry. There is a lot more to do, but all of the basics are working."

"What about the generators?"

"They're both humming smoothly."

"That would have taken ages using earthbound controllers working twenty-four-seven."

"Just the human touch. We were right when we said people were needed on Mars."

"You did all of this without *any* help from Earth Control?"

"Not a speck, but we don't want them to feel unappreciated."

They turned onto Haul Road and began rolling along a relatively smooth, straight dirt road.

"This road was winding and rutty. We're going pretty fast."

"When we sent haulers to the well, we mounted their blades to grade the roads. We also had Ivan blast out some difficult sections. About ten round trips smoothed out the road a bit."

Rob shook his head in disbelief. "Gus is going to be terribly surprised!"

"We sent a brief message through your ship as soon as we made contact. There may have been a window when the ship was still in contact with the Marscom satellite. If that worked, then Gus knows we're okay."

They pulled the four haulers into the dock area outside the air lock entrance. As they rolled inside, each was met by a shower of water that washed the collected dust away. Dust belonged outside, not inside. They all took off their pressure suits, but their friends were extremely weak and needed a place to rest.

The robots had improvised a stretcher that they used to carry the new arrivals to cots in the animal area of the Power Wing. Rob shook his head. "I'm afraid we're both very weak."

They were cleaned up by Tim and Joan, and then they were carried to their prepared beds.

Joan turned to a new food preparation robot from *Venture 2*. "You will be addressed as Martha and we expect you to respond as in human conversation. Jeb and Elsie will show you around the kitchen and the farm. We don't have an abattoir yet. We do have a few Mars-grown vegetables and some fresh rabbit meat. Could you prepare something for us to eat for lunch?"

Mary was startled. "You speak to the robot like a person."

Joan broke into a smile. "We found that they respond very well that way. We were debilitated much as you are now. When we arrived, we could only whisper requests to them and they did all the rest. We seldom give detailed instructions now."

"And you call them by familiar names?"

"We find it more convenient. We can be conversational because we no longer have the interplanetary communications delay. They each have what amounts to nametags, so you should just dive in. We treat them pretty much as humans."

Tim suggested that Rob and Mary rest a while on the cots. "We can talk when we eat."

Joan and Tim moved to the table and chairs that had been brought from the kitchen. They needed to get off of their feet. Both mixed cups of instant coffee with powdered cream.

The robots were instructed to move the new animals and plant arrivals to the farm. Most of the new animals seemed to be recovering. Jeb, Elsie, and Martha rolled them down the West Radial to the new farm. The robots set them up as the first inhabitants of that living space. Becky and Ivan began moving the plants into their space as well. Joan gave instructions.

Joan gripped Tim by his hand and whispered, "I'm still terribly worried. Rob and Mary are much worse off than we were, and we were in no great shape. Please, may we avoid these crisis experiences? They're terribly wearing."

"I only wish we could. We've only just begun. I do think we should take off a couple of days, though, to nurse them. They need TLC. We can use that time to complete setting up the farm area."

*   *   *

The global press had been spreading the word during the time out at Earth Control. The long trip to Mars had quieted the story, and the bad news of late had left the newsmen with a negative perception that permeated their stories.

Now they had a miraculous win. Four people had actually landed on

Mars and were living there. Headlines circled the globe. "Tim, Joan, Robert, Mary are on Mars. Amazing rescue. Craterwall Base is Home for People, Animals, and Plants."

At the news conference, Gus played the message over the speaker system. "This is all we know. It was just a fragment of a message that the *Venture* ships managed to get out as *Venture 2* descended into the thick dust storm. The message is terribly exciting. It tells us that Tim and Joan really created a miracle. By doing that, they were in position to rescue Rob and Mary as they were running out of oxygen. I find it miraculous that Tim and Joan could accomplish so much, even with eleven robots, most of which were really worn out. They did this in only ten days. Questions?"

"How do you feel right now?"

"Enormously relieved. These are our friends. By just going so far in space, they showed a huge amount of courage. You have no idea what this means to us. We've worked intimately as a team for over four years."

"Can you tell us what that short message tells you?"

"Tim and Joan had to work day and night while adapting to gravity again. I suspect that they developed a different working relationship with the robots. That is the only way I can see that they could have done so much. Remember that we have two more unmanned ships carrying equipment, supplies, animals, and plants that will arrive in ten and twenty days. We still must get them safely on the ground.

"This is by no means a done project. They make no mention of the mining operations or the refinery and associated operations. Their hauling road is just a rutted path going twenty-five kilometers. Furthermore, they are still in a blinding dust storm, so they have yet to see anything of the landscape."

The earth-shattering news caused a complete turnabout in the media. Everything was rosy, but there would be no more news for a while. Everyone began to speculate again.

CHAPTER 20:

KINDRED SPIRITS

TIM AND JOAN had taken turns watching after their friends overnight. They had asked their reference guide how to handle things, but that was very awkward. Joan had checked with Jeb on the use of sedatives since he handled the animals. Apparently sedatives were seldom used with animals. Finally, she had decided to give them a light sedative to help them sleep.

Joan felt their lack of direction was a problem. "Tim, we need a robot that can look after our health needs and no one thought to give us one."

"What do you suggest?"

"Crusoe. He came to Mars on the exploratory mission and became virtually useless working with the newer models. As a result, he was rebuilt as a D-Model with the latest AI and expanded memory and the latest communications. We've been too busy to do anything with him. I thought we could load him with much of Jeb's how-to and add the human stuff from the references. He could become something of a nurse and eventually take care of children."

Mary became immediately suspicious. "Are you telling me something I should know?"

"Well, maybe. I checked for pregnancy this morning and it looks like we could have a little Tim or Joan on the way. I didn't want to say anything with all that was going on."

Tim stepped over and held her gently. He smiled—really smiled—for the first time since they had arrived. "You're a wonder. I'm so lucky to have you. Love just doesn't say enough. You strengthen me and complete me."

Joan held her cheek against his and relaxed.

Tim looked deeply into her eyes. "Yes, I think we should give Crusoe a specific job, but like the others, he would do general work twenty-four hours per day."

Joan began setting Crusoe up with the software and knowledge base he needed.

Martha brought some broth for Rob and Mary. They were still very weak, but they were rested. Martha hand-fed Mary and Crusoe hand-fed Rob. Their friends were functioning well mentally, but the physical part was not well at all.

Tim sat down by Rob and Mary. "When did you stop exercising?"

Rob said, "Soon after you landed, we worked out a way with JPL to make our oxygen last. We immediately cut back on the oxygen and stopped daily activities. Twice a day, we put on oxygen masks and exercised, then reduced the oxygen again. We stopped all exercise three days before the landing."

"That was probably the only way you could survive, but you're now suffering from protracted zero G without enough exercise. You're in worse shape than we were when we landed, and that was not fun.

"This is Crusoe, the robot that came on the first mission. We upgraded him to full capability. Joan just gave him an MD using our references. He'll be your personal nurse. He knows all we have on how to deal with your problem. He knows how to treat you with physical therapy massage, light exercise, special nutrients, vitamins, and calcium tablets. We've been taking most of those ourselves to get on our feet. I'm not able to run around the block yet, so don't expect miracles."

Joan said, "Mary, I need to know something as it relates to your treatment. Are you pregnant?"

"No. We were careful, wanting to wait until we got on the ground."

"That's good because a pregnancy could complicate your treatment. With that, Crusoe surprised them. He turned to Joan and asked if he could give her a physical therapy massage. Joan laid face down on a table and Crusoe began flexing her arms and legs and massaging her muscles.

"Crusoe, why are you asking to treat me first?"

"I need a reference with a healthy person. I know you're not fully recovered, but you're the best available."

He began working on Mary after twenty minutes of working on Joan.

Joan was smiling. "You know, that wasn't half bad. I think you have another customer."

Crusoe said, "Miss Mary feels different. I'm treating her more gently. I understand that's the right way. I understand this is something that should be done at least twice a day for Miss Mary and Mr. Rob, but I think we might do it more often once I find out how well they handle it."

After another twenty minutes, he moved to Rob

Tim left Joan with their friends and headed back to the farm. What else could happen in the middle of their muddle? He called for Jeb and Elsie to join him.

"Okay, you two. Our first project is to look after all of the animals that moved in yesterday. Treat them tenderly. They're very confused. You've been giving them food by hand. You also need to have Crusoe join you and see what kind of physical therapy you can give to the animals." Tim remained with them until he saw things going as he wanted.

Joan was still with Mary and Rob when Tim returned.

Rob asked, "Wouldn't it be easier if you moved us four into the domicile?"

Joan said, "Perhaps, but we don't want to take chances. We charged that space with an oxygen atmosphere just the day before yesterday. We moved our animals into the farm late yesterday and we moved the rest in just now. Oxy is monitoring the mix of the atmosphere. We want to be sure it's under control before we expose you to any risks. This wing shares atmosphere with the plants across the corridor. This is the space we've used since we arrived and we consider this the safest place for you."

"Thanks. I should have suspected something like that. So what's your plan?"

Tim said, "Up until now, we just searched for the highest priority need and attacked it. Your arrival was our goal for getting everything in order. We almost made it. The farm was our last project. Taking care of you is now our highest priority, but getting another generator power room operating is next, so we move most of the plants to the farm tomorrow to make room for that equipment."

Mary was propped up so she could see around her. She said, "Yesterday was a bit blurry. The first part of our ride was bumpy, but the rest of the way was smoother. How much have you really done?"

"The short answer is that we did a lot of housekeeping. We wouldn't be sitting here so casually if things weren't pretty much in order within Craterwall. Outside is another matter. Right now we want to consolidate things."

*   *   *

The next morning the atmosphere within the farm was working rather well, considering it was now filled with vegetation.

There were four atmospheric regions in Craterwall: the Power Wing along the south Outer Ring, the Processing Wing along the north Outer Ring, the Habitat Wing including the Domicile and the farm on the Inner Ring, and the West Radial which tied these all together. They were continuously monitored and the air was circulated with high oxygen being mixed with low oxygen before returning to its source. It was more complicated than that because high carbon dioxide was mixed with low carbon dioxide and there were the four regions.

When all the plants and animals, including humans, were collected in the habitat, there would be virtually no life in the other regions. Then it would become a matter of stabilizing the Habitat region using the atmosphere in the other regions as a buffer. Of course, temperature and humidity were also part of the mix.

Tim and Joan were still periodically monitoring Mary and Rob although Crusoe was now standing full-time duty. The tempting smell of cooking breakfast wafted through the ventilation system, finally luring their friends from slumber. The four humans were the only ones living in the old animal room now. They all ate oatmeal and coffee together, talking between bites.

Tim broke the calm by beginning the move of the non-hydroponic plants to the farm. It went fairly quickly, followed by breaking down and moving the hydroponic lights and water races to the farm followed by transfer of the plants that depended upon the nutrient water for life support.

When the final arrangements for the people were complete, they began moving into their new home. The walls were bare and the floor was smoothed Marscrete. The overhead domes were white, reflecting light from the spiral fluorescent bulbs attached to the walls. That marked the end of their fourteenth day.

The layout had changed many times in their minds. What they settled on was placing the aquatic animals and the hydroponics in the same room. The larger animals were placed in another with the smaller animals alongside. The grains and fruits were placed in the fourth room and the vegetables in the fifth. They sacrificed human space with their beds being placed in with the kitchen and their electronic workspace. Tim made the case that the animals and plants would need the extra space when *V3* and *V4* arrived. Rob and Mary were still bedridden and Tim and Joan wanted them to feel involved with the activities while they were at their workstations.

Having Martha with them made the working kitchen area a bit of a squeeze. She had been loaded with all sorts of recipes and menus, but it took some discussion to get their personal tastes balanced.

Rob and Mary insisted that they sit at the table for breakfast. That required a bit of assistance from the robots. Still, it was a good idea to get them all on equal footing. Rob raised his hand toward Tim and said, "I suppose you have a full day of work set up for us."

Tim said mischievously, "Sure. You can each take a lap around the Inner Ring after breakfast."

Mary knew better. Neither ring was anywhere near complete, much less providing life-support. "After you."

Joan interjected a thought. "You know, we've traveled upwards of 100 million miles or 160 million kilometers to get here and we're imprisoned with only a few steps from one place to another. Ordinary commuters travel a lot more than we do. We're the great interplanetary adventurers and we work essentially at home."

Tim grunted and said, "Except for a couple of very exhilarating races through the dust storm."

Martha delivered bowls of their version of boiled ground wheat and some precious rationed coffee. Tim continued, "Actually, we'll give you a short tour on the Segways to acquaint you with our reality. It looks different in person as compared to the video we saw on Earth. We'll start with the tour that includes the processing sector and then take a break for lunch. Afterward, we'll show you the generators and how we're overhauling all of the older equipment using our new parts. Repairs include overhauling the robots and the construction equipment.

"That should be enough for today. Tomorrow Mary can watch Joan with the farm while Rob watches me work on the repairs. We use robots to work along the Outer Ring. You're going to discover, as we did, that the robots did something of a rough cut job, but they did a remarkable job of excavating and lining the habitat space after they were cut off from Earth. Everything works, more or less, but there is a lot we need to do to make it work smoothly.

"We've prioritized an immediate to-do list. The robots have unloaded your cargo into their assigned locations. *V3* and *V4* will bring more equipment and supplies that need places. We urgently need more space within Craterwall. We also need to increase the size of the farm."

Rob shook his head and said, "We need redundant access to every room for safety. You mentioned the independent Northwest Radial corridor."

Tim would not let that stand. The experience with the *Thor* generators had forced an upgrade for the new equipment. The steam turbine was twice as powerful and each of the electrical generators had twice the kilowatt generation capacity. The nuclear generator was modified to produce three times its original design. "First we set up the new generator room by the entrance and a generator room on either side of the nuclear furnace. You brought one larger generator. *V3* brings larger steam turbine and *V4* brings another of the larger generators. That new equipment provides twice our original electrical generation capacity and the increased power converts into aluminum production and Marscrete."

Joan was becoming concerned. She said, "However, that requires getting the aluminum mine into operation which requires our building a real road from the well to the mine. We can't afford to beat up our equipment on the current road like we did over the last Mars year."

Rob wanted to get the aluminum production working as soon as possible. "Even so, we really need to get back into Marscrete production. Why not divide the roadwork into that which improves the existing road and that which requires major new construction. We then select which sections are the most critical for getting us back into the mining and hauling business."

Everyone agreed.

Mary was still very weak, but said, "Don't underestimate the proliferation of our animals and plants—those we have now and those coming on the next two ships. We have little wiggle room even now. I'm assuming your old livestock room near the entrance will become a repair shop."

Tim responded, "After our scare with the generators, I really don't want to be dependent upon a single steam turbine any longer than necessary. I'll work on the new generator room until Rob is on his feet. Joan will manage the farm and Household until Mary is on her feet.

"We keep hoping for the storm to let up, but that's speculation. The haulers run on methane and oxygen from tanks. We can run them outside even in this weather. They're fairly well protected from the dust. It was the rutted road surface that got them. I suggest we go ahead and smooth out the Mine Road like we did with the Haul Road and get along with what we have. Joan can begin work on improving the road out to the mine right away. I agree with Rob that the Northwest Radial is critical but it will have to come second."

Rob offered to load the computer design for that fifteen-kilometer-stretch into the equipment memories. "I'll identify the critical work on the Mine Road. Joan can oversee the construction using Becky and Nat to physically manage the operation. They'll use explosives very generously to blow out all obstructions as you did with Haul Road while extending the new survey work on Mine Road. We're going into much rougher country as we move west. We need to straighten the road wherever practical."

Tim said, "For now, only Joan and I will take active shifts. You can watch and advise if you feel up to it. You'll double up on the Segways with us. We'll walk otherwise. We'll be working mainly from this room using our computer links to the robots and equipment.

"Mary, you can direct Elsie and Jeb on the farm for harvesting foodstuffs. Martha will preserve and store the foodstuffs. Joan will keep a personal eye on things for you and see that Jeb nudges the new animals into activity. Please do not overdo it. Just test it out today.

"I'll work with Ivan and Don. We now have fifteen robots, all in good working order. We have four haulers, two backhoe-loaders, and two mining moles that will lead the way in underground excavation. We work out task assignments at breakfast.

Mary and Rob were exhausted. They retired to the Domicile where they rested. Joan and Tim went off on work activities until noon when they all gathered for lunch.

Rob and Tim personally visited the entire constructed portion of the Outer Ring area afterward. When completed, the interior space of Craterwall would be large enough to house between fifty and a hundred people with industrial and agricultural functions.

There was no rush to strenuous labor. Mary and Rob stayed close to home after the first tour. Joan and Tim did any essential lifting and they avoided it, using robots where practical. However, all four looked forward to physical therapy with Crusoe.

Tim and Joan had quickly adapted to a different gait. They wore rubber-soled traction shoes so they would not slide and they leaned a bit more into the direction they would walk. Mary and Rob continued to shuffle with assistance from Martha.

*   *   *

Rob was working studiously at his computer practicing installation of a revision for the Craterwall Environmental Control System that he had brought on *Venture* 2. He was doing this using an independent backup CECS system on an independent computer to be safe. The CECS was an automatic system that measured the gaseous components of the atmosphere at all the critical locations inside of Craterwall and made adjustments to keep those components within safe limits. Rob had conducted this same test using a development system at JPL. Those results has been satisfactory.

Rob adjusted the proportions of the components in the backup system to monitor how the CECS handled the corrections. As expected the upgraded system applied adjustments. However that was not where it ended. The adjustments were way out of proportion to what was needed. He made manual adjustments to the CECS readings and again the response was far from what was needed. He then executed a manual override to make the corrections that were needed. The CECS system locked him out from making his changes. Rob then attempted to turn off power to the backup CECS system and again the system locked him out. Rob then unplugged the computer. That killed his computer, but it would not have worked on the primary system since that computer had a power backup to keep it running in the event of a power outage.

Rob called out, "Everyone come to my work area immediately. We have an urgent situation."

Tim, Joan and Mary all appeared immediately.

Tim kept his calm. "What's up?"

"Our CECS software upgrade was sabotaged! If I had simply applied the upgrade to the primary system it would have made spurious adjustments that would have us breathing just carbon dioxide."

"Is that the upgrade you just brought with you?"

"You're right. It even has my initials on the memory chip."

"Just a second. The backup is now corrupted. I'm going to reboot the backup computer using the original operating system that we keep in storage. Then I'm going to load the original CECS system we have been using. I'm then going to run the test data to be sure it is working properly on the backup."

That didn't work. The entire computer was corrupted.

Rob then got a spare computer and replaced all of the peripherals. He then loaded that computer from backup sources. That worked.

They all realized what had happened but did not know why.

Mary asked, "Why did it work on Earth but not here?"

Rob was furious. "Someone must have screwed up the system clock to cause a switch to the corrupted code. That or else they could sense it was one of our computers I was running on. There are many possibilities. This was not a simple computer virus."

Tim was considering what their options were now. "We're still out of radio contact with Earth Control. We need to get this to Gus as soon as possible. We could send a message upon the next landing and try to squeeze this through in the overlap gap during the landing. Meanwhile we run on the old system."

Joan was ashen. "Rob, who had contact with the revised run code or could have gotten to the revised source code?" Tim shook his head. "No one was securing the code. From now on we go to a high security footing!"

Chapter 21:

Isolation among Friends

NOW THAT THINGS were picking up and the Mars project was looking promising, their international financial supporters were taking more interest. They were ten nations, including the US, designated the G-10. Dave gave them periodic reports. Even with better reports, the oil producers were reluctant to finance the full project, even wanting to curtail the project.

*   *   *

Each day dawned with hope that the storm would abate.

By day seventeen, Tim had converted the old vegetation room beside the West Corridor entrance. That would become Generator Room B. The steam lines had been run from the stubs leading to and from the nuclear furnace to the expected steam turbine site. The new generator sat on its base ready for use. The electrical circuits were in place.

Mary and Rob progressed very slowly. Joan continued to monitor the farm activities for Mary, but focused mainly on improving the Mine Road from the well junction.

Mary was in her element. She could work directly with her robots and facilities. She purred about the animals with Jeb and repeatedly monitored the hydroponics section and other plants with Elsie. She had developed a close relationship with Martha. Mary talked with her incessantly about domestic matters. The Domicile was becoming homey. The food preparation was very creative.

Their first project was to realign the road out to the mine. A road-design program was used to create a centerline for first cut at the alignment.

The program that calculated those alignments had restrictions on how it would process the data. It ran the solution in minutes and drew the centerline on the detailed topographic maps. They were digital and would be used to guide the AI-directed equipment.

Rob directed Ivan as he staked out the centerline with light-reflecting posts even in the thick dust. Those were designed to be used with the laser guidance systems on the construction equipment. The haulers were fitted with snowplow or bulldozer-like blades to spread the soil and achieve a relatively smooth surface the full width of the roadway.

Explosives were used extensively to remove the larger boulders. Tim ended up working in the Habitat workspace. Rob was working on the road and taking over the explosives work.

Explosives are a mixture of all the chemical ingredients necessary to sustain a chemical explosive reaction. No external ingredients such as oxygen are needed. There were no restrictions on this use. Some explosives are shock-sensitive as with nitroglycerine. Dynamite may be fired using an electrical spark cap. The point is that nothing in the atmosphere is involved in firing an explosive so the same rules would apply as did in Earth's atmosphere. An explosion on Mars would be much the same as on Earth, except that an explosion on Mars would propel stones and soil farther than on Earth. Explosives are typically placed in enclosed spaces or drilled out spaces to increase the effect of the explosion.

Rob managed the two loaders cutting through winding curves and using removed material to fill empty space and to reduce steep grades. Tim seemed to love working with the explosives. The bigger the explosion, the better he liked it. He worked with Ivan in the field as he set and fired the explosives. Before long, Ivan was adept at emplacing the explosives and Tim had developed a talent for throwing the debris to fill nearby upcoming depressions.

Rob, Tim, and Joan were moving along quite rapidly. By day nineteen from first landing, they were working a full five kilometers along the new Mine Road route; only ten kilometers remained to reach the mine.

Tim and his team realized that they should not count on the dust storm lifting. They prepared another short message for Earth Control, "CECS upgrade sabotaged to make all CO2; all here okay, Go to high sec."

Early on day twenty, Tim and Joan were set up in *Venture 1* to land *Venture 3*. Joan was with Ivan and Tim rode with Becky. Rob and Mary were in the Mars Control work area in Craterwall.

Venture 3 came in high and almost out of radio range. Tim could only fix a probable landing location. He had sent corrective adjustments, but the ship was soon out of range. Tim's message had been sent for Gus at Earth control and the readings from *Venture 3* recorded.

Tim radioed Joan and Mars Control. He said, "I've calculated the projected landing spot. It's about six kilometers to the northwest—thankfully in relatively smooth terrain. Follow me."

They quickly boarded the haulers and were off, picking their way to the landing spot following a radial direction from the Omni beacon at the well.

Tim said, "I just lost the beacon and am following the compass. I hope it doesn't tumble."

He used calculated dead reckoning to trace their course on the display of the digital map. Two kilometers later, they lost radio contact with Mary and Rob at Mars Control.

After another four kilometers, Tim called and said, "I just picked up the signal from the lander and am homing."

They were slowly turning one way or another to avoid boulders when Joan called a halt.

She said, "I just spotted some fluorescence on the underside of a boulder. It was activated by my headlights. Becky will take a sample for me and mark the site so we can find it again."

Tim was a bit impatient. "Is this special?"

"Don't know. Could be."

Joan began moving again after a five-minute stop. They began gradually moving forward toward the signal.

Tim was in the lead. "I see a faint image."

The haulers picked their way through the densely scattered boulders. The ship was sitting in a clear space and level, after what had been a perfectly executed landing.

Tim and Joan were in pressure suits. They sucked in their hauler's cabin atmosphere, opened the hatches, and climbed down to the ground.

Tim said, "Here we go again." He climbed a ladder up to the hatch. There were no seats or other human accommodations. However, there were many enclosed animals and plants, including two sow pigs, catfish, trout, bees, fruit trees, blueberry, grafted lemons, and almonds.

Tim released the cargo pod, which contained the steam turbine, a loaded hauler with a robot, and a variety of metal-shaping machine tools. He activated the new robot, which joined Ivan and Becky in unloading everything onto the other haulers.

They began the tricky part of driving the haulers back to the well, using their gyroscopic compasses. They took a southeasterly route from the lander until they picked up the Omni at the well and could contact Mars Control.

Mary at Craterwall was quick to respond. "Glad to have you back in contact. How was *Venture 3*?"

Joan answered, "Perfect condition. Everything is loaded and we're on the way home in all three haulers. There was a recording of a message from Earth Control."

He played the message for all to hear.

"This is Gus Hoover speaking for the entire Earth Control team and everyone who contributed to your journey to Mars and the facilities you found upon your arrival. Part of your message was received from the *Venture 2* landing. We take it that all four of you are now safely on Mars and both Mary and Rob made it okay. Tim and Joan, we find it most remarkable that you could do so much at Craterwall on your own. Mary and Rob, we know your life signs were strong upon your arrival so we expect that Tim and Joan have taken good care of you. We congratulate you all on this unprecedented grandest achievement for all mankind.

"Everyone on Earth is stunned by all four of you being on Mars. You are now historical figures. Virtually everyone around the world has been celebrating.

"What follows are messages from world leaders. We look forward to reestablishing radio communications with you and hearing much more of what you are doing."

Joan picked up the communication to base. "Everything is secure and we're just now reaching the well. We're turning the guidance of the haulers over to the robots."

*   *   *

The Earth Control team was waiting for any message from Mars that might follow the *Venture 3* landing. It was played on the loudspeakers and into the electronic media receivers.

"All okay. Mary and Rob are recovering from extended zero G, lack of exercise, and oxygen deprivation. Haul Road and five kilometers of Mine Road improved. Farm okay. Mine Road and Northwest Radial next. Love to our families. Thanks to you all. Greetings to everyone. The Mars Team."

*   *   *

The haulers zipped down the Omni chain to the Craterwall entrance in just twenty minutes.

The robots unloaded their cargo. The corridors were becoming packed. The four humans held a lunchtime conference. Tim started by acknowledging his responsibility to install the new steam turbine. That would give them a half backup for their electrical power generation. Rob and Mary were worn out and would take a break. Joan was ready to continue work on the Mine Road.

Tim said, "Joan, What was that sample you picked up near the landing site?"

"That was a gift from on high. You know the solvent we use in the aluminum smelters to dissolve aluminum oxide in the ore? That originates from a mineral called fluorite. It absorbs incident light then emits its own light in a fluorescent process. Our headlights caused that to happen in the sample. That stuff is quite rare and we just found a deposit. Smile, you just received a supreme gift."

Everyone was a bit stunned; it was far more important than jewels or gold, particularly on Mars.

Tim reviewed their perspective. He said, "When *V4* arrives, we just won't have room for everything. We can squeeze the bio stuff in the West Radial corridor, the Domicile, and the farm. The other generator can go in that room, but the reset of the machinery and supplies must stay in the dock area just outside the entrance."

Rob said, "We should be able to reach the mine before the *V4* landing. In fact, I expect we can even go to the mine and begin starting up the operations. If that is so, we should begin hauling ore to the refinery about that same time. I've checked out the smelter operation. We should have it ready to take refined aluminum oxide in another couple of days."

Tim said, "I should be able to install the *Venture 4* generator about the same time and devote that entire power source to the smelters. We will then need to increase the capacity of the refinery and add more smelter pots."

Rob said, "When we complete the Mine Road, we can bring some of that equipment back to begin carving out the Northwest Radial. A loader and a mole must operate the mine. The haulers will be carrying the ore back here to Craterwall."

Joan said, "Perhaps by then the sun will come out."

On the twenty-third day, the first batch of forty-eight baby chicks hatched. Mary was beside herself. She and her robots watched over them night and day. The chicks were born in Mars gravity and took to it naturally.

By the twenty-fifth day, the road was within five kilometers of the mine.

Chapter 22:

Earth Contact

THE TWENTY-SIXTH DAY was their day for celebration. That morning, a bright spot appeared in the sky. They could actually see a bit around the robots on the ground. It was not like the end of a rainstorm. The dust in the air just slowly thinned out.

By late afternoon, the alarm alerted the team that connection had been made with Marscom. A deluge of queued messages flew from Marscom to Craterwall and from Craterwall to Marscom and Earth.

It was night outside. Tim sent the first live message from Mars with its growing transmission delay as Earth outran Mars in their orbital race.

"Gus and all, you have no idea how refreshing it is to have communications back. First, did you get our message during the V3 landing? Extremely important.

"Everything is okay here. Rob and Mary are on limited duty, but slowly recovering. Our rooms and corridors are jam-packed. The farm is overloaded and we have rabbits in our sleeping quarters, which are in the kitchen with our Mars Control workspace. We set up the new steam turbine so we have real backup for half of our power.

"We're hugging and kissing each other in a frenzy right now and only wish we could hug each of you, too. You just don't know how empty we've felt without you on the other end of the line. You're our tie to reality and to our home. We've not seen anything but dust and the bare walls of our corridors and rooms for almost four weeks. Joan, Mary, Rob, and I are jabbering to each other about what to say next.

"Thanks for the ferrets and finches. They have certainly enlightened our lives. Each of us has a pet ferret and they've been sleeping in our bedroom, but they seem to be nesting in the farm. The finches are not yet hatched, but we will place them in the rooted plant area where they can fly about.

"Oh, yes. We're within just a few kilometers of the mine and plan to send in robots tomorrow to check things out. The dim sunlight is most welcome, but the dust still prevents us from seeing very far. We need the sun because we'll use solar power to get the mine up and running. The road is improved and when completed in the next couple of days, we'll be able to haul at fifty kilometers per hour. We'll be hauling from the mine to Craterwall in under an hour—so we'll be moving ore far faster than last year.

"Thanks to every one of you for all you have given to make this possible. You're our lifeline. We've endured some very difficult times, but we could not have survived without all you have given us to work with."

The time delay scrambled the message. Fortunately, the digital-to-voice translation included the tone and twang of the original speaker's voice.

"This is Joan, Mary, Rob, and Tim ... the Mars Team."

Concurrently, a message arrived from Earth.

"This is Gus. What is the story about the CECS revision? We tested our copy and it works fine. Our deepest heartfelt greetings to all of you on Mars."

Another message read, "I see a message from you coming in as I speak."

Gus continued, "We are terribly excited that we can now communicate directly. You will never know how relieved we were to receive that first piece of a message you got back to us when *Venture 2* landed. We didn't know if you would see the opportunity.

"We marvel at how you've pulled so many things together with so little to work with, particularly in those first ten days. We know you were not in the physical condition to do much after all of that time in zero G. We hope that you're taking good care of yourselves.

"We just now received your second message. You're working on the right things. The power was first, then the mine, and then the Northwest Radial. You must not ignore the farm, but Mary is right on track with that.

"I now see that the *Venture 3* landed okay and that you got all of the cargo safely to the base. We know that you are very crowded for now.

"We sent greetings from your families with all of the communications you must now be receiving.

"The world is hanging on everything you tell us. It's quite hard for most people to accept that you are now on Mars. You are real life science fiction.

"I just saw that you found fluorite near the *Venture 3* landing site. Joan, that is a terribly important discovery.

"We'll catch up and then get back to you.

"With the warmest respect possible from all of your friends on our team and around the world … Gus Hoover"

"Gus, the CECS problem is extreme. Someone got hold of our revision and inserted a poison pill. In short, it locked us onto the program and made changes that would make out environmental atmosphee all cabon dioxide. Fortunately I was testing the upgrade on a backup system and was finally able to kill it by unplugging the power. We set the computer in a do not touch place. It has a mind of its own. We are still running on the old system," Rob

Gus read the message over and over. Every person and animal at Craterwall would have died if Rob had run the upgrade on the primary system. Gus' immediate reaction was to call in his security staff and the technical team that had worked on the upgrade. Their first action was to isolate the upgrade programming and to search its contents on a seeker program. Sure enough, their copy of the run program had the spurious code embedded. Rob, you need to destroy the upgrade copy and the system where you tested it. We will find out who and how this happened. We are also implementing security precautions to be sure this cannot happen again." Much more passed back and forth as the people at Earth Control and JPL caught up and analyzed what had been going on.

The words that Tim and Joan had said for history had been sent in the first burst of recorded messages. What remained was the human aspect of space exploration. Tim and Joan had not been in contact with Earth Control since landing. It seemed much longer than three weeks.

Gus came back on line and said, "Our prayers have been answered! Wow. We're all celebrating. You're all okay and everything's working. What a miracle!

"We have a lot of work to do. Our first Mars year is mainly to expand and stabilize our base of operations and begin reproducing animals and birthing children on Mars, as you know."

Joan said, "Mars is our home from here on in and the sooner we learn its character, the more we can accomplish. We're very eager to get out on the surface, but right now, we're extremely happy to have a secure place on a very dangerous world.

"We've rebuilt and upgraded the original robot from the exploratory mission. We call him Crusoe. He is now our medical support staff, monitoring and caring for Rob and Mary with physical therapy. He's a godsend. One more piece of news. Tim and I want you to know that I am now about six weeks pregnant. Crusoe will help me when the time comes."

Everyone laughed. Everything was not just okay—it was great.

Gus said, "Very hearty congratulations from us all on the upcoming first genuine Martian. You're okay with some modest exploration. Sure, take your time. After all you have been through, there is no reason to tempt fate. You are our future. Please be careful as you tunnel. And turn over everything that is not urgent to your colleagues here.

"Word of your success has been passed along to the global media. You've sent us an extraordinary story that has everyone excited. The reporters are asking for all sorts of information. We'll collect their questions and pass them along for your response when you feel up to it. Please remember though that this is pretty much the biggest media story ever."

Tim said, "Tell the media that we send greetings to all the people of Earth expressing our appreciation for this wonderful experience. We're settled in our excavated habitat. We have yet to see any distance so we have yet to see Mars as a landscape.

"Please send us only reasonable questions from the media. You will now receive messages from our team for their families and loved ones. I'm sure they wondered how things were going here, too. Please tell my mom and dad that we did well and that I love them dearly. I'll be sending a message after the others."

With that, the longest of long distance messages were sent between those who really cared.

Renewed communications opened a floodgate of messages and technical queries and reports. The press wanted to know about their health. Earth Control had all of their physical monitor data, but what they really wanted to hear was how it felt to be on Mars. They told the story of the landings and their journeys in the dust storms. They told of the animals and plants. Art and Gwen, the ferrets, became household friends of all the children launching a flood of ferret purchases and many stuffed animals. Then Gwen gave birth to a litter of kits.

The twenty-seventh day was much clearer, but still a bit dusty. The twenty-eighth day was bright and sunny. The team put on their pressure suits, rode out onto the open road, and took a walk around to get the feel for the real Mars.

Joan was looking back at the Genesis Crater. "You know, it's larger than I expected and much higher."

Rob laughed and said, "It's billions of years old and has a twin in Arizona. But we've made it very different."

Rob dug the toe of his boot into the ground. He had brought a baseball outside. "Tim, catch."

Tim saw it coming and caught the easy toss. They each moved back a few steps and threw the ball again. Rob jumped a bit to catch it, finding that he went higher than expected. He landed off balance and nearly fell. "Kind of floats, doesn't it?"

Joan had a soccer ball. Mary was down the road a bit. Joan gave the ball a little tap and it scooted down the relatively smooth surface.

Mary caught it with her foot, and then gave it a little pop into the air. Joan headed it back to Mary with her helmet. Mary caught the ball and dropped it to the ground to send a grounder back to Joan.

The light horseplay stopped as they realized that Rob and Mary were still limited. They all had enjoyed the feel of outdoors movement in the light gravity—even in their pressure suits.

*   *   *

Tim and Ivan completed the last of the explosive excavations. Joan and Becky cleared a rough path down into the mine area. Tim had Ivan check out the abandoned open pit mine. From Craterwall, Rob directed Ivan to improve the access and organized the mine activities.

On the morning of Day 31, Tim and Joan were in *Venture 1* to assist in the landing of *Venture 4*. The ship came in about fifteen kilometers west of the landing target, but apparently made a good landing. The radio contact was good in the clear air, but the intervening terrain was very rugged. They used the radial direction from the well Omni beacon and *Venture 4*'s measured distance to the beacon to locate the ship. The livestock and plants on the ship made rapid recovery essential.

Rob used the detailed maps from the *Thor* mission to pinpoint the exact landing site—only to find it directly on the new road at the approach to the mine. The Mine Road was definitely the best route. Tim and Joan headed out the new higher-speed road—a road that had not existed in its entirety until a day earlier.

Rob was amazed. He said, "Hey, do you guys believe in a divine intervention? This is just a bit too coincidental."

"Don't say anything until we get there."

Joan was in the lead and Tim was close behind. They were running sixty kilometers per hour, pushing the safe speed just a little. They approached the designated spot tracking on the radio beacon of the lander. They made the last turn more slowly only to find the ship sitting in the very middle of the road—a spot that the day before had been a large rocky prominence.

Joan stopped just short of the ship. "This is just a bit too spooky!"

Tim pulled up beside Joan and gave the electronic order for the ship to lower its cargo pod. Meanwhile, Becky and Ivan opened the hatches and began transferring the life-support containers to the haulers and connected the hauler air hoses.

Tim reported back to base that everything was in order. "The only problem is in moving the ship so we can access the mine."

With that, the three haulers began the drive back to base at a more acceptable speed. Their friends met them as they cleared the entrance air lock. All four looked at each other in wonderment.

Rob finally set it to rest. He said, "I'm not about to question this good fortune. It just happened."

Chapter 23:

Missing Hurts

Everything had changed. All of the landers were safely on the ground and the cargo from *Venture 4* was stowed in Craterwall. The new equipment was installed and, most importantly, the plants and animals were settled in. The roads to the well and the mine were completed in gravel. Tim and Ivan installed the second generator. Rob had the refining operation ready for bauxite and the smelting pots were ready to accept alumina. They were ready to begin hauling bauxite.

Mary and Rob mounted an excursion to the idle bauxite mine. They set out before dawn with two haulers. Rob rode in the life-support cab of the first hauler and Mary rode behind.

Rob looked around at the scenery as they prepared to move out. He noticed three bright stars in advance of the rising sun. The one nearest the sun was a pale white. The second was brilliant white. The third was a double star: one larger and blue white, the other smaller and the same pale gray color as the star nearest the sun but not as bright. He realized that he was seeing Earth from the surface of Mars for the first time. He also noticed faint vapor clouds on the eastern horizon that disappeared almost as soon as the rays of the sun touched them. The clouds seemed terribly foreign over the desert landscape.

Rob commented to his fellows who joined him in the experience. Seeing the quadruple morning stars was enough to calm him down a bit. There was home. There was the image of the sun's immediate planetary family in a single grouping. He and Mary were traveling alone across an exceedingly foreign landscape.

"Rob, you know we've spent all of our time inside our habitat. This view seems surreal, like I'm living a dream."

He said, "It's real, but still, it's not. We have each other, but everything else seems out of place."

The current journey across the surface continued with subdued excitement. Tim and Joan had journeyed through the dust storm to rescue their incoming ships and even to rescue the living things on *Venture 4*, but this was Rob and Mary's first personal exploration. It was the first journey across the foreign landscape. Seeing Earth from this perspective left them feeling a long way from home.

The most obvious difference on the surface of Mars was the color and the occasional silhouettes of impact craters. Everything was covered with pale pink sky and darker surfaces—not the familiar blue sky and green landscape on Earth, and that was true regardless of which direction they may look. The terrain was mildly rugged everywhere. Larger craters could be seen in the distance along with a bluff where the land rose significantly in the west. The most prominent impression of the land was of open, empty expanse in all directions until it disappeared behind a relatively near horizon. The curvature of Mars' surface was twice as great as Earth.

The sun was noticeably smaller than when seen from Earth. The sky was sunset red near the horizon since the rays passed through residual dust from the last dust storm. But the sky was surprisingly clear and the sun was more of its normally orange Earth tone when looking high above the horizon. The impression was of a day on Earth that was slightly hazy from air pollution.

Mary and Rob could see Deimos and Phobos—two small natural satellites—moving slowly across the heavens. They spotted them from time to time moving very slowly across the sky, one eastward and the other westward. One orbited faster than Mars rotated and the more distant orbited more slowly, but still advancing relative to the celestial sphere. Their Marscom and GPS satellites were in geostationary orbits, closer than one moon and farther than the other, remaining directly over the equator.

The vehicles passed the well by early morning and continued on to reach the mine at midmorning. They made a leisurely tour of the mining site and the refinery operation. Their first excursion in their spacesuits was to explore the mine diggings. The old solar panel electrical generator was uncovered and inspected. It was in working condition. Rob turned it on after using hauler power to orient the panels toward the sun.

Meanwhile, the backhoe and bucket loader began filling the haulers with ore. Three haulers were standing by to take loads. They would carry the ore to a prepared vat at the Craterwall refinery.

At Craterwall, Tim and Joan would see to mixing the bauxite with solvents and hot water. The solution flowed into a heated, high-pressure stone container where the aluminum oxide dissolved out of the bauxite. That liquid was transferred to a settling vat where the alumina precipitated out of the solution. It was washed with fresh water to cleanse it of chemicals, and then reheated to dry the damp white powder, which was the alumina end product. The process was relatively involved. The residue spoil was collected by the bucket loaders and hauled to a nearby depression where it was dumped with the spoil from the previous year.

By late afternoon, they settled in for the night. All of the construction equipment carried oversized generators so they could be used as electrical power sources for onsite power equipment. The power systems of the two rovers were coupled for safety and Rob transferred into Mary's vehicle. At sunset, they each ate the space version of an MRE, reclined their seats, and settled in for a night away from home. They called Craterwall and described the remarkably brilliant red sunset that lasted only briefly then disappeared. All was well and they fell asleep.

Rob and Mary were working with Ivan and Becky for two more days to get the mine operations up to speed. Finally, they told Craterwall that everything was in order and left the operation in the hands of Earth Control, the robots, and earthmoving equipment as they headed back to Craterwall. By that time, the operation was averaging two loads an hour back to Craterwall and getting ahead of the refinery processing operations.

Rob and Mary sped home on the freshly constructed roadbed, covering the distance in just thirty minutes. They had shrunk the apparent size of their operations dramatically. The haulers were checked for needed repairs at Craterwall and immediately sent back to the refinery for more alumina. By day's end, they had hauled fifteen relatively small loads of alumina to Craterwall and were quickly loading the electrical pots to begin actual aluminum production.

Having all four of the new ships safely on the surface was grounds for celebration on Mars and on Earth. All of the new equipment, livestock, and vegetation had been carried inside of Craterwall. Gus could finally breathe a sigh of relief and face his superiors and the world press at ease. They had crossed the Rubicon of space, placing humans safely on another world with all they needed to survive for at least four more Earth-years—two Mars years— long enough for two more fleets of four interplanetary ships to reinforce Craterwall at the next and the following planetary oppositions.

So much had happened in just six weeks. Everyone on Mars and most on Earth considered these results to be the grandest human achievements of all time. Congratulations streamed across space, commending the astro-colonists for their accomplishments. The four on Mars felt mixed emotions at the moment ... relief of course, but also a sense of an enormous burden with long workdays ahead and a prayer that the future would not strike them with challenges that they could not manage. Still, they felt reassured and confident. They had faced and overcome enormous difficulties. They had moved into their new home and the pieces were coming together.

* * *

A reserve of alumina accumulated at the smelter. The electricity from the small nuclear power plant was turned on to the pots. It was now a two-pot operation in keeping with the growing operation at the mine and the increased available power.

Direct current electricity flowed through the solution and aluminum powder settled to the bottom of the pot. The process was turned over to Earth Control becoming a routine operation.

Aluminum and the Marscrete mixture were becoming abundant. The Marscrete allowed priority to be shifted toward excavation of the Northwest Radial corridor. The pure aluminum allowed forming aluminum sheets, bars, angle irons, and most particularly pylons to carry generated power and the fiber optic data lines directly to the mine terminal.

The most critical operation was extension of the corridors and addition of more rooms within Craterwall. The AI robots, vehicles, and equipment excavated the Northwest Radial to the Inner Ring. They moved counterclockwise, "southerly," along the Inner Ring to join with the habitat section. The team would then move backward along the Inner Ring constructing rooms.

When completed, it provided an emergency exit from the farm and, as rooms were constructed, it also expanded the domicile space for the farm and the home. The construction was handled by the Earth Control team, except for the interior work needed to prepare the space for habitation. That completion required sealing the corridors and rooms, pressure-testing the sealed space, and adding the environmental equipment. That work was directed by the Mars Team.

The twenty-four new rooms on the inner corridor were put to use. While one construction team went to work on the Outer Ring, another began upgrading the roads with a sturdy base.

In the past three months, the animal and plant population had exploded. The entire West Radial corridor was packed with animal and plant life. Ten of the new rooms were immediately occupied by the overflow, including incubation, seedling growing, veterinary, grafting, and potting. Four more were an expansion of the human living space. Of the five such rooms, two were mini suites for the couples, one was an eat-in kitchen, one was the shower and utility room, and one was the Mars Control work area. Five became duplicates of the processing system rooms on the Outer Ring. That gave them processing redundancy for distilling water, producing essential gases, compounding methane, and mixing environmental gases.

The Mars Team built the processing equipment using complex parts brought from Earth and simple pipes and metal structures produced on Mars. One room was used for human recreation space and another for a human medical facility. Three were used for equipment repair, manufacturing facilities, and a machine shop. All of the space was occupied almost immediately.

The information was tracked in their breakfast review meetings, except one meeting was different. Joan was showing signs of her pregnancy, including morning sickness. Her condition was becoming a matter of concern for the entire group. Tim mentioned that Joan needed to take occasional rest breaks during these spells.

Mary took this occasion to make her own announcement. "Rob and I need to give you some news on our part. I tested pregnant myself about three weeks ago. You have two pregnant women to deal with now."

That completely changed the discussion of the moment. Tim immediately said, "Mary, we couldn't be happier. Are you doing okay?"

"We performed all the normal tests. Everything seems just fine."

Rob wrapped his arms around his wife. "We decided to hold off in letting you know. Too much was happening around here. I expect we should include this information in our next message to Gus. We also want to let the grandparents-to-be know."

The expecting couples continued work within Craterwall. Joan was now four months pregnant and it turned out that Mary was actually two months.

She focused intently on the animals and plants. Rob and Tim set up small-scale textile equipment in one corner of their domicile. The rooted plants now included cotton. A first, if modest, crop would be ready to provide fibers for cotton thread and weaving of cloth diapers for the babies.

Gus and the Earth Team were becoming more and more aware that the scope of the Mars operation within Craterwall was not to be determined by the number of colonists, but by the very diverse production operations needed to supply their growing needs.

Work on upgrading the primary roads was nearing completion. When equipment became available between other jobs, it was used to build a firm stone base for Haul Road and Mine Road. Upgrading the base required screening equipment that was designed to selectively screen various sizes of rocks from the spoil unearthed during the roadwork.

The team continued to work hard on creating their new world from the world they had once known on Earth, all in the hope of making new discoveries. And it so happened that a big one was just over the horizon.

Chapter 24:

When Least Expected

THREE WELLS WERE in operation. One was at the original exploratory landing site, one was where Haul Road met the Mine Road, and the other was at the bauxite mine itself. Each of these was enclosed in glass much like hothouses used by nurseries on Earth. The design captured and contained as much heat as possible from the distant, weaker sunlight. The heat was needed to provide protection against the pipes freezing and thus reduce the need for electrical heating. Since it was not a pressure-tight space, the inside air was originally the same pressure and the same chemical composition as the natural Mars atmosphere outside.

Joan and Tim frequently explored outside for natural resources, going out as much as fifty kilometers. On one of these excursions, they entered the Haul Road well's glass enclosure. A small amount of blue-green caught Joan's eye, then she realized it was all over the place. She scraped off some samples for Mary to identify back at Craterwall. She also took a sample of the enclosed air. They were headed up to the first landing site so they looked into that well enclosure. They found more of the blue-green material there.

Tim and Joan were both a bit excited with their find, and Mary was speculating along with her two friends.

"If we were on Earth, I would consider a fungus, algae, or moss to be candidates. I just don't know what it could be here."

And so it remained until the explorers entered their living space at Craterwall. Mary and Joan made quick business of identification once they had their magnifying and analytical equipment to guide them.

Joan smiled broadly at Mary.

Mary said, "It's not foreign. We have some of it in the air here where we live. We brought it from Earth. It's cyanobacteria. What does surprise me is finding it growing on its own at barely 1 percent of Earth's atmospheric pressure.

"Cyanobacteria is thought to be one of the first life forms to develop on Earth. It's primitive and very hardy."

Joan completed the puzzle. "The atmospheric content within the two well enclosures has much more oxygen and less carbon dioxide than outside."

Mary said, "We've provided a humid, warm environment with plenty of carbon dioxide. Sunlight is in abundance during the day. The cyanobacteria consume the carbon dioxide and give off the oxygen. All Mars needed was the moisture from the well and some seed to kick-start life growth here. We have ultraviolet light to kill stray allergens in our Craterwall atmosphere. We better report this to Gus and JPL."

Tim made the contact as soon as they could organize their findings and their suggestions. The distance between Mars and Earth continued to grow as Earth outdistanced Mars in the orbital race. The lag in radio transmission was growing as well, by about eleven minutes.

Earth Control, this is for Director Gus Hoover and for any further disposition only by him.

We have a discovery inside the well enclosures. Those enclosures were intended to improve heat retention at the well sites. The initial enclosed atmosphere was Mars standard. Because of the function of the well, we inadvertently also captured moisture within the enclosure.

Cyanobacteria were found growing rather abundantly within the enclosure, most predominantly inside the glass where the sunlight was incident. We also found that the enclosed atmosphere was no longer nearly all carbon dioxide but ranged around 20 percent oxygen. It appears the Mars standard atmosphere was infiltrating the enclosure with carbon dioxide stabilizing at approximately the 80 percent level. The cyanobacteria appears to have been converting the incoming carbon dioxide to oxygen and ingesting the carbon to grow more cyanobacteria.

For perspective, cyanobacteria have been around for billions of years on Earth. As you know, it's credited with converting Earth's early substantial carbon dioxide atmosphere to include over 20 percent oxygen, a very corrosive chemical. That's how our home world came to have animals and plants, the one converting oxygen to carbon dioxide and the other converting carbon dioxide to oxygen.

That's all happening in our well enclosures at approximately 1 percent Earth standard atmospheric pressure. We had not anticipated that cyanobacteria was carried here by our project or that it could grow at that low pressure. However, we have no indication that it will grow without the substantial presence of moisture.

We note that, despite the low atmospheric pressure on Mars, the partial pressure of carbon dioxide here is actually not much less than the partial pressure of carbon dioxide in Earth's atmosphere. That is a rather remarkable coincidence … or is it?

Consequently, this opens the possibility that some varieties of carbon dioxide-dependent plant life could grow on Mars if enough warmth, natural sunlight, and moisture were provided … at normal Mars atmospheric pressure. Pressure-containing structures may not be needed. At the least, we anticipate we could produce breathable air using cyanobacteria. The question is how to proceed.

The Mars Team

Then they waited the double eleven minutes for a response … and then more. Finally, the response came directly from Gus.

To the Mars Team,

Congratulations upon your discovery. We always hoped for some exciting native Mars discovery, but your discovery is something else.

The biological team here is beside itself. Our question is whether we can bio-engineer more complex plants that can produce food for animals at Mars atmospheric pressures. If so, we can use outside thermo-pane hothouses and greatly expand your farm operation. The food we grow may be something you never imagined. Give us a little while to digest this. You have created a radical turn of events.

Gus Hoover

The next morning, they found a message from the botanical component of Earth Control.

To Mary,

You have the real thing there to do some broadcast testing, more immediately available and not simulated as would be necessary by us. You discovered some plants will grow so we will just test variations to see what the plants themselves like best. What you found is remarkable, but we need to find out just how far we can push your discovery.

Construct an enclosure similar to what you have around the wells—except, in this case, use double-pane glass to keep in the most solar heat. Use the natural Mars atmosphere, but watch its composition. Remember to try all your variations—even Spanish moss. It grows off of air and moisture, too.

The Mars colonists were becoming quite excited, hoping that they could grow some useful plants outside. They did not expect to be able to grow plants outside in winter, but they did want to be ready when the next growing season arrived.

Meanwhile, some of the surface dust had been magnetically enriched to the point that it became high-grade iron ore. Pure iron was produced using electrical processes ... but the four generators had reached their electricity generation capacity. Some alloying elements were being found, but usually not in rich ores. Many other chemicals were found, but not in rich deposits.

The explorers occasionally came upon exceptional mineral finds that had long ago been depleted on Earth. One ancient wash turned up gold nuggets as well as obvious flecks of gold in the wash bed. On Mars, gold made excellent contacts for electrical equipment. Tim and Joan collected bags full of the nuggets purely for fun. When they returned to the habitat, they shared the nuggets with their companions. Mary suggested that the gold might make an interesting substitute for the weights that they carried upon their bodies to keep them in shape. Rob worked the nuggets into bracelets, anklets, belts and headbands, giving them all an exotic appearance.

*   *   *

Joan and Tim were able to send their humbee flyer out for some serious exploration with emphasis on exploring the three fossae clusters just north of Craterwall. Their first step was to map the nearest fossa in detail. Once they had 3-D color imaging superimposed on contour maps, they could begin plotting geological information.

Ever since exploring the Craterwall site, they had known that there was low-level radiation in the area. It would not threaten the living things, but it was still there. Even the water they had brought up in the wells left residue that was slightly radioactive.

Finally, they were in position to descend physically into the nearest fossa. That fossa was over fifty kilometers long, as much as eight kilometers wide, and 350 meters deep with steep walls and heavy talus debris fields at their base. Tim cut a descending trail into the face of the most easterly terminus wall at a place where it was largely free of talus debris. A cut was made through the remaining boulders, stones, and silt debris to the center of the depression.

Geothermal features in Yellowstone Park supported microenvironments year-round; extraordinary colonies of small plants and animals flourished despite the surrounding deep freeze. The geothermally warmed fossae lakes on Mars could have been locations where life on Mars might have evolved and survived longer than anywhere else. Joan directed the robots to collect samples from the fossa bed in dozens of scattered locations for later detailed inspection.

Stratified sedimentary deposits visible in the walls of the fossa included an outcropping of bauxite, probably an extension of the mined strata. The robots doing this work collected strata samples from the talus. The material was catalogued by discovery location in sample bags and brought to Craterwall for Joan to analyze later.

Joan had taken the opportunity before leaving Earth to carefully inspect the famous four-and-a-half-billion-year-old ALH-84001 meteorite sample from Antarctica. The heavily researched stone showed evidence of what might be very primitive bacterial life on very early Mars. The Mars origin was determined from gases the meteorite contained that matched the chemical composition of the modern Mars atmosphere. Joan checked her stones of like composition for similar content.

Joan did have the ability to determine the age of stones by dating of radioactive decay of even slightly radioactive trace elements found in the stones. She always made a first check of her samples for such dating. Her measurements determined that virtually all of the stones in the talus pile were terribly ancient—over four billion years old—and native to Mars.

The surface rocks were another matter. The geological history of Mars was very different from Earth. The process of accretion continued from residual clouds of meteors that encircled the sun in those days. As time passed, the accretion did not stop, it only slowed down.

Building the roads on Mars had unearthed many meteorites from the softer soils. They told an extraordinary story of the accretion process over billions of years. Joan's collection contained surface rocks from the fossa, stones from the talus piles, and surface rocks from the road building.

Joan passed the gist of her preliminary findings along to her colleagues on Earth, but out of necessity, delayed most of her work until the winter hibernation period.

She had another collection that sparked her interest. Each of their three wells brought up brine loaded with other impurities that required settling before distillation to obtain pure water. Joan had many samples of those briny waters for analysis.

One morning, she became excited when the Geiger counter detected high radioactivity in one of her samples. It tested to contain uranium. She found others also contained very rich uranium. She excitedly called for the others to meet her in the nursery where Tim was working while minding the children.

The lack of an abundant supply of fuel for power generation was the most serious problem facing the colonists. Perhaps rich uranium ore could yield pure uranium and that could be enriched in the radioactive isotope, something of a long shot. Joan made a rudimentary test for volume of the sample. She compared the source to the amount of radioactivity to make a rough estimate of the amount of uranium in the ore. Everyone became excited. Tim wanted some idea of where Joan's discovery might lead.

Rob said, "All I remember from basics is that the process of enrichment involves a gaseous compound of uranium that can be spun very fast in a centrifuge, forcing the non-radioactive and denser U-238 gas to the outside, which displaces the lighter and radioactive U-235 gas to the inside. At least that was the theory. The problem was that only a very small fraction of the U-235 moved toward the inside so the process had to be repeated in a series of these centrifuges. Besides, we need to determine if this discovery is an isolated anomaly or if it's substantial."

Tim said, "Rob, can you chase this down with references we have on hand? I really do not want to pass this along to Earth Control until we know better what our choices may be. Joan, could you pin down the locations where those samples came from? I'll then see what else may lie in that area."

The four explorers were working by radio in their workspace when Tim dispatched a flyer deep into the fossa. Joan was watching Tim's monitor over his shoulder.

The floor along the base of the walls was loaded with boulders, so they moved out into the middle where the original floor was flatter and less littered. They found a few open spaces with smaller stones and sediments. The radiation counter had been turned on with only mildly interesting counts. As the ship approached open space, the counter began to hum with activity, actually going off the search-sensitive scale, which automatically switched to less sensitivity.

Joan struck Tim in the shoulder.

"Ouch. I agree," said Tim.

Rob and Mary joined them. Tim directed the flyer around the edge of the sports-field-size open space. Still the counter was humming continuously.

Mary said, "The whole area is alive with high radiation. Let's get a sample to analyze!"

Tim shook his head. He said, "There are spots like this all over the central floor. I want to check the others."

With that, he raised the humbee and moved it to a smaller nearby space. The counter just slowed a bit as he moved it, but it kept up its buzz. The intensity increased again as he approached and flew over the new site.

They were all very excited. Tim needed some explanation. He said, "Okay, Joan. Give."

Joan smiled and said, "This is very hot—much more than I ever imagined. There are a number of radioactive elements, but the one that comes to mind is uranium. Radium and thorium are possibilities, but their half life is too short. This has been lying here for an eternity. The pure uranium oxide is only about 0.7 percent U-235, which is the radioactive isotope. The question is what percent of the ore is uranium oxide. We really need some samples to test. This blows the mind away. This could make it possible for us to produce our own uranium fuel rods to generate electricity."

CHAPTER 25:

STARDUST

TIM WAS LIKE a hound on a scent. "This could give us all the nuclear fuel we dream of."

Joan saw the difficulty. She said, "A miracle. We would need a lot of very sophisticated equipment that you could not carry inside our ships. Besides, Gus made it very clear that our sponsoring nations really did not want us to have that ability. We can enrich uranium ourselves. I've enriched uranium, but we must first have pure uranium metal.

"We could refine the ore into pure uranium oxide and then reduce that to uranium metal. The next step is to chemically convert the uranium into uranium hexa-flouride or UF_6, which is a gas. The gas is commercially enriched in U-235 on Earth in a series of centrifuges. The centrifuges spin the gas so that the heavier U-238 gas tends to go to the outside and the U-235 gas moves to the inside, but the difference in mass is very small. The spinning is very fast—centrifugal forces are extreme on the spinning chamber, tolerances are very tight, and getting the separated gases out of the spinning chamber is very tricky. The process is an engineering nightmare."

"No hope?"

"We need a lot of help from some very experienced and expert people. That cannot happen unless Gus and the G-10 nations want us to do it."

Tim had gathered sample soils and had their little flyer on the way back to Craterwall. He said, "We should have the very small samples here in about ten minutes. Joan, after you do the analysis, we need to send a report to Gus. We should get busy figuring out just how to make our case."

Joan was busy setting up for the testing. What were the ore components and what was the percentage mixture? Joan took her time, being very careful with the dangerous chemicals and to be exact in her measurements.

Joan said, "There is some covering dust. The ore itself is a mixture of uranium oxide or U_3O_8 with typical Mars detritus. The richest grade of ore on Earth is about 15 percent U_3O_8 and the U-235 isotope is only about 0.7 percent of that. The grade of uranium oxide in the ore here appears to be significantly higher than that. Our isotope abundance is the same as that on Earth, since all uranium would have accreted in the formation of Earth and Mars at the same time and the decay rate would be the same. Uranium's half life is extremely long—about as long as Mars is old.

"The refining process would produce more uranium metal here because our ore is richer. The chemical formation of UF_6 would be the same as Earth, remembering we already have fluorite for our aluminum processing. The centrifuge process involving molecular mass would be just as difficult here."

Joan paused and then said, "There is something I should tell you. My resume does not show that I worked on uranium enrichment. I'm very familiar with the process. I can design and build a centrifuge facility to enrich the uranium ore we have here. The general concept is an oversimplification. The challenge is in the precision technology and techniques."

That was completely unexpected to the others.

Rob cleared his throat to get their attention. "I also have something to tell you. I worked on plutonium breeder reactors. We can design a reactor that will convert the uranium 238, which is not radioactive, into plutonium—and plutonium is radioactive, making it a fuel much like uranium 235. There are tradeoffs depending on the reactor design and fuel ratios. There are complications and risks as well. I know how to manage those complex design choices."

Tim said, "I should reveal some more of my own background. I worked on the team developing a practical plasma rocket engine for the US Air Force in a security classified project. The main advantage of a plasma rocket is that it can provide propulsion, however small, for the entire duration of an interplanetary voyage. The engine design was successfully tested in a rocket that circumnavigated the moon and back. It would have been capable of reaching Mars in only one month. I know how we could build such a plasma rocket engine here. That design also involves some very advanced technology."

Everyone was now watching Mary very expectantly. Would she also have a technological miracle?

Mary smiled tentatively and said, "I spent over a month at Ames Research Center before we launched. What interested me the most was the new discoveries in growing human tissues, organs, and even embryos. I can now grow normal people in vitro as a woman can in her womb. We can gestate embryos in artificial wombs, resulting in children as normal as Joan and I are growing in our uteruses. That process is very tricky and moral concerns limited how far researchers were willing to go. We could radically increase our population growth using the frozen human eggs and the sperm we brought with us."

It changed everything. Tim expressed his concern. "I don't know of any discussion of this technology with any NASA personnel. Gus was selecting us for this mission. Our having this technical expertise is too coincidental to have happened by chance. I expect Gus wanted us to have this knowledge. I also expect few—if any—others know we have this expertise. What do you think?"

They were thinking over the ramifications. Mary responded. "We and Gus are the only ones that are likely to have assembled this information into a whole. He has not mentioned it, probably because the others responsible for our project would not want us to have and use this kind of technology. I think we absolutely must keep this information secret among ourselves and should require unanimous agreement before we use any of the technology."

Tim asked for an open discussion of the ramifications.

Rob said, "We can manufacture all the nuclear fuel processing equipment we may want. A breeder reactor converts U-238, which is not radioactive, into plutonium, which is radioactive and a heat-producing fuel. Once we design and build breeder nuclear power plants, our effective supply of fuel will be increased a hundredfold.

"With the plasma engine, we will not be limited in space transportation. Interplanetary travel becomes much more economical and much quicker because the low-thrust plasma engine continues to operate during the entire voyage.

"With the artificial womb, we can reproduce as prolifically as our rabbits do—and our women need not be encumbered with pregnancies beyond those they may choose to have. Our limitation will be in nursery care and parental nurturing of children and their education."

Mary said, "Please realize that the kinds of things I am offering are often regulated or prohibited on Earth. Also Tim's plasma engine is probably still classified. We can't make any mention of these developments in our contacts with Gus and Earth Control. I don't think we should mention this to the newcomers on the next ships—at least not until they are settled."

All four team members participated in writing and rewriting the message about the uranium discovery. Finally, they agreed to report the essential facts of their discovery and let the cards fall where they may with no mention of their newly revealed abilities. The ramifications were obvious.

Confidential

To Earth Control, Attention Director Gus Hoover:

This morning, Tim and Joan made a major discovery of numerous very rich uranium ore deposits in the fossa near Craterwall. Joan analyzed the samples. The ore is about 23 percent uranium oxide and the isotope ratio is about the same as is common on Earth.

We all feel very excited about this discovery and hope you regard it the same.

The Mars Team

The reaction was explosive when the message reached Earth Control. Gus had been playing down a deteriorating political situation on Earth. The G-10 industrial nations were very confrontational. Terrorism by the Shiite faction of the Arab world was getting out of hand and oil to the rest of the world from the Persian Gulf Region was direly threatened. The price of crude oil was extreme and the flow of oil was greatly reduced. Nuclear power plants were springing up everywhere, but native nuclear fuel was becoming expensive as well.

Global warming was also becoming evident. Oceans were rising with the melting polar ice packs and violent tropical storms were becoming more common. The global economy was on the verge of collapse.

Consequently, support for the Mars project was virtually nonexistent. Gus had been warned that funds were to be cut off after the next two fleets were sent. Gus had known of Joan's expertise—and the others as well—but he was keeping that to himself. He was desperate to get his colony self-sufficient right away and he needed to help Joan begin an enrichment project now that he knew about the abundant uranium at their doorstep. Without fuel for the nuclear power plants, the colony would die—literally.

To the Mars Team:

Your message is momentous. Please confirm immediately with details. You must give us some time to work out an appropriate response. Joan has something to offer now that you have uranium ore available.

Gus Hoover, Mars Project Director

Confidential

To Earth Control, Attention Director Gus Hoover:

The floor of the proximate fossa is virtually covered with radioactive locations. We can explore the deposits only by flyer. We will be using improvised explosive projectiles to send larger explosives to the floor and thus will excavate deeper depressions.

We shall begin developing a vehicular route down into the fossa, but that will take weeks—if not months—to complete since it is through rugged, boulder-strewn, steeply sloping terrain.

Joan is the one who made the discovery and is very excited. Is there any prospect that we could enrich nuclear fuel here?

The Mars Team

There, they had said it. Gus read the message over and over. He did not want to tell their international advisory board of the find until he knew if it was substantial. A substantial find could change an awful lot. Earth was very hungry for fuel and Mars could become part of the answer. At best it was a gamble.

To the Mars Team:

You are getting ahead of matters—both here and on Mars. This must *not* become public. The global situation here is very sensitive.

Even if you had unlimited support, you could not conceivably construct an enrichment centrifuge in the near term. You must think in terms of what you actually control. You must explore and map your find. It could be very meager or abundant. Do not jump to conclusions.

Gus Hoover, Mars Project Director

Their immediate objective was to construct a vehicular roadway down the sloping wall of the fossa. Meanwhile, Tim and Joan flew the flyer over about a quarter of the floor nearest to the terminus of the access roadway, mapping its radioactivity in some detail. The radiation extended across the entire section with variations in intensity. A substantial, deep surface layer of non-radioactive material would have to be laid at the road's terminus to provide mass to absorb the emissions from the floor. It would become a safe working space on the floor for machinery.

The next exercise was for the aircraft to fly a grid across the entire floor. The radiation again varied in intensity, but it extended along the entire length and width of the floor. The fossa did not end at the western end, but turned northerly. They did not seek to map that further section or the two other fossa that lay a bit farther north.

An explosive projectile was launched into the canyon and directed to explode more or less in the middle. The resulting crater was five meters deep and twenty meters wide at the top. Numerous samples were taken and they showed continuing radiation at all levels and similar composition. Even if the radioactive strata were only five meters deep, the expanse was so great that the ore was very abundant—more than enough to justify mining and refining operations.

The road down the eastern face was being constructed in haste so that they could respond promptly to Gus. Explosives were used extensively. The idea was to construct only so much of the working platform as was essential in order to set up the drilling rig on the platform. The drill would be used to extract cores going down as far as the radiation continued to be present in the core samples. The resulting map and specimen analysis were transmitted to Gus. Then the team waited.

Gus had been more than busy with his own geological team and engineering experts flown in from the recent American Centrifuge project. The reaction was universal. The measurements must be faulty. Such a rich deposit was beyond any deposit ever found on Earth. Gus's team was placed in direct contact with the Mars Team and every method imaginable was used to expose failings in the Mars measurements. Videos had been taken of every step of the process and exhaustive documentation and images were transferred across the enormous distance between the two planets. No one wanted to believe the measurements despite verifications in every imaginable dimension.

CHAPTER 26:

POWER AND PERSISTENCE

TIM LED THEIR discussion on rocketry. He said, "This is a two-step project. If we can launch a chemically fueled rocket into Mars orbit, it will be a big deal. We can do that using our old lander rocket nozzles and fuel tanks we assemble here. We can fuel it with liquid methane and liquid oxygen we produce here. I can design and build a small rocket to practice launching into orbit.

"But what we really need to design and build is an interplanetary chemical rocket that can carry a payload of unique Mars specimens back to Earth's orbit. We need to send samples using a conventional chemical rocket. We can do that because we can send a normal chemical rockets up from our low gravity base all the way to Earth's orbit. Conceptually we use only about a quarter of our fuel for the basic launch and have an appreciable amount of fuel remaining to maneuver the rocket into an interplanetary braking maneuver, putting the rocket on an elliptical solar trajectory back to Earth. It will be increasing velocity as it reaches Earth when another rocket burn must put the rocket into Earth orbit. NASA must then transfer the payload to an earthbound shuttle.

"Until now, we've used chemical rockets that gain their thrust from the momentum of the burned exhaust gases focused through a nozzle. The mass of the fuel that would become the exhaust gases is a burden to carry up into space inside of the rocket.

"NASA began an exploratory project to develop a different technology that is radical. This is the concept. If an ionized material were magnetically accelerated to a point near the speed of light and ejected from a magnetic nozzle, the velocity of the expelled ions is exceedingly high and the ionized mass that must be carried inside the rocket is quite small."

Rob and Mary needed some clarification.

Tim continued, "Getting up into planetary orbit and descending from planetary orbit requires the powerful propulsive force of a chemical rocket. Traveling from Earth's planetary orbit to Mars' planetary orbit does not require so much propulsive force. For that purpose, the low propulsive force from a plasma rocket can gradually raise and lower a plasma-rocket-propelled ship far more efficiently.

"Dr. Franklin Chang-Diaz, an experienced astronaut, began work on a plasma rocket while employed at NASA. He retired from NASA and formed his own company, Ad Astra Rocket Company, in January 2005 to continue development work on a Variable Specific Impulse Magnetoplasma Rocket, known by its acronym, VASIMR. As you know, his company recently tested his VASIMR rocket in a space environment.

"When we launch from Mars, we can carry more than enough fuel. We would end up in orbit with over 75 percent of the normal fuel load available for return landing and the rest available mass for launch-landing cargo. Until we have plasma, we can use the excess chemical fuel to carry cargo to Earth's orbit, albeit on a much longer trip than with ultimate plasma. We could also refuel from one chemical rocket to another using EVA robotics, supplementing with fuel needed for that interplanetary trip. We have choices now."

Rob was catching on to the enormous potential Tim had revealed. He said, "This sounds like a lot more than a way to ship uranium to Earth. If you take the obvious next steps, Mars becomes a big dog in the space business. That is a lot more than anyone here or at NASA had in mind for us. It also means the powers on Earth will be seeing us as competition in a huge way. I really think we should be very careful how we play our cards. We need to keep a very low profile for as long as possible. We must do everything possible in secret—just among ourselves."

Mary and Joan were stunned. Mary said, "If you combine this with our other secret knowledge, this becomes explosive."

Joan was very anxious. She said, "I think we are going to be confronted with unwanted attention. I think we need to develop a plan. We should focus on our own uranium enrichment plans for our own sake. Send only samples of our innocent discoveries and just park it in orbit."

*   *   *

They quickly moved on with construction of Rob's chemical rocket design. The engine was tested and found to be just what Rob had anticipated. They did not want to show any more of their cards than necessary.

The interplanetary cargo ships would carry samples of natural uranium metal—just to show what they were starting with. Earth Control would have the responsibility of getting uranium metal from orbit to the surface.

The project was set in motion to build a rocket to test the launch capability. This activity was not disclosed to Gus or Earth Control.

The first challenge would be in constructing the cargo shuttles on Mars. Tim and Rob decided that a demonstration was the best way to prove their point. They had the rockets and fuel tanks from the landers for propulsion. The landers were designed for non-destructive disassembly so components could be separated. They drafted four robots to assemble the first rocket from the lander components.

The rocket was ready for ground tests in less than a month and an improvised launch facility ready by the same time. The individual rockets in the powering cluster were separately fueled with liquid methane and liquid oxygen and the engines were tested. The control systems were also tested for electronic performance and communication with Mars Control. The only payload for the Mars to orbit test was a communications and control package.

Early the next morning, the fuel tanks were topped off with liquid methane and liquid oxygen. The team was ready for launch. Tim sent a message to Earth Control.

> Earth Control:
>
> Be advised that the Mars team has reassembled and tested a rocket vehicle and will launch that rocket into Mars orbit immediately. We will send you real-time launch countdown and launch performance beginning now. Please monitor for further messages.
>
> The Mars Team

They picked up the launch count at five minutes. The count progressed smoothly, the nozzles of the single-stage rocket gimbaled to launch alignment, and all four rockets in the cluster fired simultaneously. The ship rose rather briskly from the launch pad, reached maneuver altitude and velocity, and then inclined easterly.

It was about this time that the message reached Earth Control and an alert was sent to Gus and the others. Gus was not one to like surprises, particularly this kind. He was beginning to realize that his Mars Team was entrepreneurial—a characteristic he had wanted them to display, but not with these surprises. He knew what Tim wanted to accomplish and he sympathized. A success here would demonstrate that the Craterwall base was in the space business as well as the Mars surface-exploration business.

The launch was being reported over the Earth Control loudspeaker for everyone to hear. The rocket engine performed flawlessly and the climb to orbital elevation and velocity was by the book. The rocket's engine shut off and tests were conducted onboard to determine if the ship was actually in orbit as intended.

The verbal message from Mars recommenced:

Earth Control:

All indications are that we have a successful launch and orbital insertion. We're picking up GPS signals that show we are in a 300-kilometer-high orbit and stable.

The Mars Team, Your Partners

Gus was in the control room by then. He just put his head in his hands. He had planned a much more paced project, culminating the following opposition with a return ship. Gus was now pushed ahead into Mars-to-Earth rocketry. All he could do was send a message of congratulations. However, launching a rocket into orbit actually gave him the leverage to maintain the Mars project at an even faster pace—and to push for a uranium centrifuge enrichment project on Mars. Now he had to design ships and an entirely new project to bring the enriched uranium back to Earth.

The report of the successful orbital launch had gone out immediately to the world and was played back over the electronic global media. Finding the Mars Team could launch space vehicles was a big surprise and generally received with elation.

The Earth's populace had previously seen the Mars project as a big adventure of questionable value. Now it was seen as science fiction coming true. The G-10 council was angry that they had been left out of the process; even Gus's denials of personal involvement did not ease the tension. Still, these were the big industrial powers in the world and finding that an alternative source of fuel might be mined and enriched on Mars was most encouraging.

Another aspect was that—with Mars gaining unlimited fuel—their population and activity could be greatly expanded. It created an illusion of

Mars becoming a power unto itself. The prospect of importing significant quantities of urgently needed nuclear fuel was the dominant perception.

It all seemed remote to those involved. There were only four colonists on Mars and the prospect for the upcoming opposition was to add only six more. Furthermore, the Mars productivity was being driven by robotics and those were being controlled by JPL and Earth Control. It was very far from the science fiction depictions of future life on Mars.

Developing the needed technology on Mars was certain to prove daunting. One prospect to move things along was the use of carbon fiber technology. High tensile strength carbon fibers had been commercially developed in the last quarter century. The latest designs of uranium enrichment centrifuges had been of carbon filament design, providing very light weight and very strong drums for the centrifuges, allowing higher spinning rates.

Tim and Rob had planned a bit ahead in regard to carbon filament technology since Mars did have that carbon dioxide atmosphere. Rob and Mary began tests on forming carbon filaments and forming the carbon filament fabrics and laminates.

What they needed was the technology to form precision cylinders that they could use as fuel tanks for returning rockets and the centrifuge drums for the uranium enrichment. That was not regulated and they had that information. They also needed the centrifuge designs that were tightly held by the Oak Ridge National Laboratory and the U.S. National Regulatory Commission.

They did have one advantage over Earth's uranium enrichment. Uranium hexafluoride is highly toxic when mixed in the presence of moisture. Mars has about the driest environment one could imagine.

Meanwhile, the construction within the Craterwall ring was continuing. When the Outer Ring between the West Radial and the Northwest Radial was completed, they used that space to add another twenty-four rooms. It was allocated entirely to manufacturing, equipment repair, and industrial processing, including operations moved from the Inner Ring. That freed up space on the Inner Ring and allowed more space for agriculture.

The released space was reallocated with sheep and goats while pigs, chickens, rabbits, and fish each had separate rooms. Three rooms were set aside for hydroponics and four more for soil growth of plants.

The increase in space was much more than utilitarian—it resulted in relief of the sense of confinement the team had endured for the previous eight months. The earlier major accomplishments had been technical, but this was human. They declared it a day of Thanksgiving. On Mars and Earth, the entire NASA and JPL team celebrated the event. Many people on Earth joined in—it marked the realization that people and living things were really established on Mars. With the other discoveries and activities outside of Craterwall, the entire base was seen as an astounding success.

The team decided that it was overdue for some incidental road building. The first project was to construct a road around the perimeter of the Genesis Crater, soon named Perimeter Road. While they had many photographs of the back side of their home, they had never actually seen the view. Next, they found a more sloping, spiral rise up the crater wall to its crest, one they could climb in a loader. They constructed a gravel road along the entire ridge line of the crater.

With that in place, they climbed aboard a pair of loaders and took the scenic ride. The steep drop into the bottom of the crater was almost two hundred meters, which was below the level of the outside surface. Still they were at 150 meters along the Crest Road, which gave them an impressive view of the surrounding terrain and across the crater to the other side. The view of the interior of the crater bore a distinct resemblance to Meteor Crater.

The view to the east was of the boulder-strewn expanse of the Acidalia Planetia, the ancient seabed of the enormous ancient northern Mars Ocean. To the northwest were three enormous sunken fossae; the nearest was close enough for them to see down into part of the interior. To the west lay the distant foothills of the Tharsis Bulge where the aluminum mine was located. To the south lay relatively rugged terrain with many drainage slews out of the bulge leading easterly to the Acidalia Planetia.

The Mars calendar was offset from Earth's. The best way to keep track is to start with the date when Tim and Joan first landed—six Earth months since they launched. Joan had become pregnant just before landing on Mars so she was due for delivery not quite nine months from that time and Mary was due about three weeks later. That made Joan's due date fifteen Earth months after the Viking launch. A Mars year is about twenty-five Earth months long. Earth will catch up to Mars in opposition about twenty-three months after the *Viking* launch or nine months after Joan's expected delivery.

Joan began her final month before birthing her child. All the preparations were made in the dispensary and the child care facilities were in place. The circumstances changed their mental perspective in everything they were doing. The women gave prenatal examinations to each other following what they had learned by observing and assisting with actual deliveries. They were very experienced midwives.

Since equipment was being used continuously, the maintenance facilities were expanded and used extensively. The machine shop was being used more and more to construct new equipment and manufacture parts for the old ones. Meanwhile, a new electronics laboratory was completed and equipment installed. They began producing electronic chips, boards, and computer memories as well as power supplies and other electronic components. Chemical production facilities were being expanded daily. Newly discovered natural resources were used to gain everything from everyday replacement supplies to conversion of gaseous nitrogen into fertilizers.

All of this was being accomplished with heavy support from Earth Control, but there was a limit to progress because of the limited resources. Most critical was the current supply of enriched uranium or potentially enriched plutonium. They kept an eye on fuel utilization.

The four colonists had become even more committed to their work. Physically, all four people were gaining strength, bone mass, and dexterity. They were all carrying ballast weight, bringing their combined weight to full normal Earth weight.

CHAPTER 27:

MIRACLE OF BIRTH

JOAN WAS EIGHT and a half Earth-months pregnant and Mary was just three weeks behind her. The birthing room was of standard size; however, a separate area was set aside for other medical procedures.

The third and fourth ships had brought two goats and two sheep that had matured before launch. They had come into heat. They had been artificially inseminated with frozen sperm upon gaining their footing on Mars. They had already delivered their kids and lambs. The mothers were most important as a source of milk—not only for the adult human diet, but also to supplement human mother's milk as soon as the women could deliver human offspring.

Progress was exploding their facilities and the real-time effort of their human workers on Mars was a large part of that progress. They set new projects in motion and then turned them over to the controllers on Earth. The current interior of Craterwall would be fully utilized as living and working space before the next ships could arrive.

Joan reached term in her pregnancy just thirty-four weeks after the landing. Her anticipated childbirth had been followed closely by all of those who had been stepping aside and assisting her over the past few weeks. Mary was close behind and too awkwardly extended to make Joan's delivery. Both women had been in various degrees of physical and hormonal distress during the trying, historic weeks.

Mary was experienced with recent birthing, albeit with animals. She had brought along only the basic surgical implements, medical equipment, and supplies. Even those occupied more space and added mass beyond what they had originally planned. Joan did bring frozen blood supplies from Earth matching the blood types of each team member. Each of the couples had been tested for compatible blood to avoid complications with their offspring.

The care of pregnant women, the care of infant children, and delivery of babies had been anticipated during their training. Their DNA had been tested against genetic defects as part of their applicant screening. Their thorough physicals had even tested the men and women for fertility and the women for adequate pelvic capacity for normal delivery.

Over the past four weeks, Mary had been giving her comrades instruction in how to help her with the delivery should the need arise. Tim would serve as her assistant. He learned what to do in the event of complications, particularly for a baby that might be positioned for a breech birth or even a Cesarean section. They joked that using Rob as a midwife would likely lead to an engineering design for the child. There was one prospect that they held in reserve. Would Crusoe be called into action?

It was general public knowledge on Earth that both women were in advanced stages of pregnancy. The press reported a countdown to each of the anticipated delivery dates, particularly for Joan since she was expected to have the honor of delivering the first child on Mars.

Tim was hovering over Joan to the point of restricting her daily activity. The afternoon that she felt the first contractions, she kept them to herself. She told Mary when she was confident that they were not the false contractions that she had occasionally experienced. The others became suspicious when Mary disappeared into the dispensary to check that everything was in order.

Sanitation was always a concern—their sanitation procedures for the delivery were primarily for cleanliness since the entire living space was considered clear of dangerous microbes.

When Joan's interval between contractions reached five minutes, they moved into the delivery room. Tim was allowed to comfort Joan, but was only allowed to watch the delivery from a distance. Rob was standing ready. Mary had prepared Joan, showing Rob what she was doing at each step.

"Honey, just keep pushing. I can see the top of the head. It's going great … stay with me now. We won't be long. Come on. Here we go."

The infant was cradled within Mary's hands. It seemed terribly small. Mary placed the child in a cloth held by Tim and quickly snipped and closed the umbilical cord.

"Joan, you have a beautiful little boy."

Everything was happening quickly. Mary cleared the child's breathing passages and gave it light pats to be sure it was breathing well, checking that the child had no obvious abnormalities. She applied drops to his extremely sensitive eyes to protect them in the oxygen-rich environment.

Joan had the exhausted look of a marathon runner. She smiled weakly as

Tim brought the child over and handed him to his mother in a small blanket. Tim bent over and kissed Joan and the first human born on Mars.

"He's beautiful. We agreed to name a boy John after your father," Tim said.

Everyone cooed and fussed over the child until Mary pushed them aside. She placed John in an aluminum version of a typical nursery bucket under a warming light. He would be watched day and night for the next few days. Mary would sleep nearby and Crusoe stood close at hand.

Tim, as leader and father, issued the message announcing John's birth to those at Earth Control. "Tim Randall and Joan Wall Randall proudly announce the birth of their son, John Randall, at 17:50 hours July 42, 56 (local time and date) at Craterwall on the planet Mars. Mother and child are doing very well. John is a perfectly normal and healthy child. Photos of John and his parents are attached."

There was a dramatic new definition to normalcy on Mars. The colonists were needed in their professional duties. Still, everyone somehow found time to include John as they moved back and forth. Two days later, Joan was picking up more and more of her normal duties. The entire family was helping with John since Mary was still "mother" to all of the plant and animal life.

The remainder of the Inner Ring was completed with rooms between the two radials during this period. A decision was made to extend the ring across the two radials enough to add six additional rooms to each end.

Mary was watching her own pregnant condition. All tests and indications were positive for a healthy child, but no one was certain. Joan kept a close eye on her friend's condition. Mary went into labor a week early and the experienced team sprung into action. Mary's labor was short and delivery came almost too quickly.

Joan announced that the baby was a girl. She performed a cursory physical check for normalcy, breathing, and other indicators, and quickly applied the eye drops. Tim took the child and proceeded with her cleanup. Joan finished with Mary, keeping an eye on her life signs. Mary announced that she and Rob had agreed to name their daughter Elizabeth.

Rob delivered the news to the waiting world that little Elizabeth had joined their company as the sixth human on Mars. It turned out that Elizabeth did not fare well with nursing. Mary decided to try goat milk in its place. It worked much better and a crisis was resolved.

The children were sleeping in a crib in the parents' bedrooms. Mothers took care of the overnight feeding, but the fathers took turns as nursery parents during the day, slowing projects down a bit.

Three weeks later, construction was completed in Craterwall. The carbon dioxide atmosphere was pumped to a near vacuum and then the oxygen atmosphere was pumped in to replace it. No one trusted the new space entirely for the first week, but finally Rob declared the new space ready for occupation. The air locks were opened along the Inner Ring, the Middle Ring at the West Radial, and along the Outer Ring.

Suddenly, the occupancy space within Craterwall had more than doubled. In this case though, there was no rush to occupy it. Everyone had come to realize that they had been fortunate to arrive at this expanse without major crisis. There was a double check on all environmental systems, particularly environmental testing equipment. They were intent on discovering any irregularities at the very earliest moment. They also were active in adding not only redundancies, but a third level as well.

Time seemed to fly. With only five more months until the Craterwall was placed in hibernation for the winter, the emphasis was on doing everything possible outside before they were confined to the base's interior.

All the useful natural resources were being acquired, processed, and stored within the empty rooms along the Outer Ring. Distilled water was frozen and stored deep underground where the temperature was always far below freezing. Abundant supplies of oxygen, nitrogen, and methane were stored in high-pressure cylinders, also in underground spaces.

Haul Road was compacted with a top layer of smaller stones and then sealed with thick Marscrete. The Space Port tarmac was spread with stone as was the connection to Haul Road and then both were sealed with Marscrete. The Port Road ran the distance from the original human landing site at the well to the Space Port tarmac.

All ships except *Thor 3* were moved to and positioned on the tarmac. *Thor 3* remained at its landing site. A large area west of the tarmac was cleared of larger boulders as a preferred landing site. Radio and laser guidance systems were installed to facilitate the next spring's anticipated landings. Equipment was installed along Crest Road and tested. The colonists had learned from their own landings in the dust storm. They were now prepared to handle anything short of the most electronically obstructing storm without leaving the confines of Craterwall.

Rob turned over direction of much of this outside work to those at Earth Control. He personally undertook manufacture of replacement equipment with primary focus on steam turbines and electrical generators. He had managed with guidance from Earth to construct a steam turbine except for the highly stressed impeller blades that required special alloys.

He had been able to construct all of the electrical generators except the armatures—bundles of fine copper wire wrapped continuously around a shaft. Elecromotive force or voltage was produced by spinning the armature in an intense magnetic field. The armatures required special materials and precision winding of myriad fine wires.

Likewise for the nuclear furnace, he could make everything except the very close tolerances and controls needed to insert and retract the fuel rods that produced various amounts of heat in the furnace and steam.

Drilling equipment was constructed to sink wells of larger pipe diameter. More armatures for larger electric motors needed to torque the drilling process would be brought from Earth. The basic methods for this manufacturing were proven.

John and Elizabeth were normal babies—with one exception. They were early in turning over. This concerned their parents because their bones were not yet strong enough for strenuous movement. In fact, a host of potential problems with growing children faced the colonists. The children were also given golden bands to wear, much like their parents. They were eight months old by the time Craterwall was closed for the winter.

All of the circular rooms in the habitat space contained a pit in the center of the floor for rich soil. Small fruit and nut tree saplings were planted in each room. The trees included citrus fruit, apples, apricots, almond, and cherry.

Some of the older vegetation space adjacent to the human habitat was being converted into living space for the newcomers. The dispensary was becoming a small hospital.

The two completed radials with their connecting ring segments now formed an interior corridor loop. It quickly became an exercise track for walking and jogging. A competition soon developed to decorate the loop with paintings.

The additional industrial and manufacturing space was being set up for electronics manufacture, metalworking, woodworking, fabric weaving, knitting, clothing manufacturing, and more. One room was set up as an abattoir for slaughter of animals and processing of their meat.

The oncoming intense cold compelled completion of the Southwest Radial and the connecting Outer Ring and Inner Ring segments. The corridors and rooms were being lined with Marscrete blocks as they were excavated.

The glass-enclosed greenhouses were evacuated. Mary had pushed very hard with her research into new plants that might grow in the extremely low pressure atmosphere. She was successful with some leafy plants such as grasses, lettuce, and spinach.

The idea of a winter confinement was mainly conceptual. The colonists had been spending most of their personal time in Craterwall anyway. All of the operational activities outside, including wells and mines, had been closed down. All equipment had been brought inside. There would be no further construction or remote excursions.

A final report was sent to Earth Control confirming a checklist of activities. The outside shutdown status was accomplished. Everyone seemed to relax now that there were no further outside risks. All radio links were working with the Marscom satellite. Their new companions would launch in three months and arrive in nine months. The hibernation could end in as little as eight months.

Chapter 28:

Life, Liberty, Pursuit

THE ORIGINAL MARS team, with intense support from Earth, had accomplished far more than anyone had expected. The Craterwall base was expected to be ready for all the newcomers as well as house more plants, animals, and much more equipment. It would be occupying more than a quarter of the capacity of the crater for arched corridors and domed rooms by the time the fleet arrived. The incoming mission was named Pioneer. Much more expansion was planned over the next Mars year.

Plans were greatly increased in physical and technical goals. Those included constructing uranium enrichment centrifuges, mining uranium ore, refining uranium metal, producing uranium hexafluoride gas for enrichment, constructing nuclear fuel rods, and constructing interplanetary transport capacity to ship a demonstration load of nuclear fuel rods at the next planetary opposition. The transports would carry all that was needed to construct more than two dozen AI robots and perhaps local manufacture of far more.

In order to accomplish these ambitious goals, the composition of the interplanetary fleet had been changed from four ships carrying people each to six ships with only two ships carrying people. New equipment, many repair parts, and robots were at a premium. The two couples that would proceed to Mars upon the upcoming conjunction had been selected—or so they thought.

With the success of the first colonists, a hue and cry arose from other G-10 nations wishing to physically participate in the project. All of the expected Americans were pushed aside to make room and two more crew members were added. The politicians felt that it was better to include representatives of their nations in the project to establish their physical control on Mars.

The Mars Team discovered this turn of events in a message from President Sorley and NASA Administrator Dave Bagnal.

> To the Mars Team:
>
> Your discoveries have caused quite a stir here. The United States and the G-10 nations have decided to greatly increase support of the Mars Project. We will cooperatively increase the number of ships at this planetary opposition from four to six and will increase the number of new members for your team to six people with much-needed expertise. The new members will be selected by the other G-10 nations.
>
> We will provide you with uranium enrichment technology, allowing you to expand your power generation capacity and providing the fuel needed to make you self-sufficient. We will also provide the means for you to construct space shuttles and interplanetary transports from Mars, enabling you to rapidly become an exporter of enriched uranium for Earth.
>
> We look forward to your small group growing rapidly into a strong energy resource for planet Earth.
>
> President Sorley

The news was not entirely welcome on Mars. In fact, it set off a firestorm of discussion. The team members immediately turned to each other. Tim moved them to the dining area; their dining table doubled as a conference table.

Tim said, "Okay, each one of us should give a first impression of what is motivating the change in our reinforcements. What is likely to be the outcome? We must work rapidly. A delay in response could be taken as a negative reaction from us."

Joan was quick to respond. She said, "I suspect that the economic and terror crisis is much worse than we have been led to believe. You just do not gain such rapid reaction from ten powerful international governments if they are not experiencing a major crisis."

Rob said, "Please also remember who we are dealing with. These people are adept at diplomatic negotiation. We are not. They are also very powerful people who hold our future in their hands. They are going to be reading our response very carefully."

Mary nodded and said, "Right now, we are four people who see our world in much the same way. Bringing in a diverse international group would likely upset our consensus-building relationship and inject more politics. Six new people would certainly give the newcomers the ability to push our opinions aside. They would almost certainly be selected as much for their political views as their technical expertise. The G-10 could well want us to become blind followers of their policies and decisions. In other words, we become their workers with little real say."

The team was quickly and strongly going on the defensive. The problem was how to avoid giving that impression to the powers back on Earth. They needed to find out more from Gus by moving their communication to him—and they needed to quickly establish a rationale that would protect their own interests.

Tim said, "Let's first express our appreciation for their generous increase in support, and then immediately shift our response to be through Gus. Since they see us as technicians, they would expect us to be responding in that vein. We might also point out that the greater part of the work here is done by robots and not by people. We can ask for more robots and fewer people, pointing out the high cost of transporting people and our great need for sensitive equipment.

"We might also direct our technical issues to specific individuals with strong credentials, steering away from individuals with national ties that could draw us into conflict with the competing elements on Earth. We don't ask what they have in mind. We give them a very positive project plan of our creation; we recommend new members we know we can work with; and we use our own credibility while we still have it."

The Mars team had been working so closely together for the past six years that they could almost finish each other's thoughts. The draft reply was ready in barely an hour. The review and final draft was ready in only fifteen more minutes. That speed almost set a record for a quick response.

Gus knew his team very well. When he saw the response was directed to him, he smiled broadly. They had read the situation well. It was almost telepathy. They asked for only four new colonists. Two were from the NASA/JPL control teams. Ho Huang was third-generation Chinese-American and expert in carbon fiber laminate technology. Norika Timoshenko was second-generation Japanese-American and an expert in enriching uranium for nuclear power plant fuel. One of the other two was Josie Hardwick, a British expert in electronic systems manufacturing, including robotics and artificial intelligence. Arnold Wilson was an Australian expert in mining and refining uranium ores. They were viewable as broadening the diversity of the Mars Team. Those missing were Moslems, Jews, and others with ties to the explosive Middle East or to terror-related conduct.

Gus made a few minor changes in the Mars Team project plan and sent it on to Dave, the White House, and then to the G-10 board.

The board was not happy with the number and makeup of the additional colonists and even insisted upon changes that were clearly political. Their plan was forwarded through President Sorley, Dave, and Gus to Mars.

The Mars Team responded in positive fashion so far as was possible, and then made a strong point that the makeup of the additional team members would not be able to achieve the goals being set by the G-10 board. They pointed to their own experience in setting up the very successful base and urged that their views be given strong endorsement.

Dave, the NASA administrator, shot back a response that the Mars Team was not being asked to approve the G-10 plan, but to simply execute it. Dave did that to crowd the Mars Team into action—and he got it.

The Mars Team fired a reply back to Dave. They made clear reference to their special technological skills that could see them through if the G-10 chose to be obstructive. Each member signed the response, stating that they endorsed the goals of the G-10 plan, but organization and execution must be left to the discretion of the Mars Team. The G-10 board members simply did not have the experience to effectively and intelligently deal with matters on Mars.

They urged President Sorley to support them in their position. As a postscript, they pointed out that they lived on another world where politically focused concerns had no bearing upon survival and success. If introduced, they could only hurt. Dave and Gus endorsed the response in detail and forwarded it to President Sorley, but not to the G-10 board. They knew the mettle of the Mars Team. The team was far more than a group of technicians—they were pioneers who had clearly proven themselves.

President Sorley called in his representative on the G-10 board as well as Dave and Gus. He said, "I appreciate that we are on the short end of a four to six balance on the board. Our people on Mars could see through what was happening here. I respect them for what they've done—and how they've pulled miracles out of thin air. They were reminding us that they could build a plasma rocket and construct a breeder reactor—among other things—if necessary.

"We've tried accommodation, but the Mars Team is right. They are accustomed to assessing very difficult situations and taking risks when they become necessary. We can't afford to let the G-10 appeasement majority take power on Mars. We can do no less than follow the example of Tim and his colleagues. The G-10 board must accept the current Mars Team leadership or else we cannot provide Craterwall with resources for their use. The US has the expertise and resources to go it alone if necessary. Carry that message back to the G-10!

"There is one concern. The Mars Team sees this as making them independent of our crisis. I think it puts them square on our side—and vulnerable to the terrorists if the terrorists can penetrate their team or stow something dangerous on those ships we are sending to Mars.

"Can we count on the loyalty of all four of those that the Mars Team selected to join them? One or two would not shift a decision, but now we are putting each Mars Team member in a position where they will be handling secret information. Do they pass the test?"

Dave said, "Since you put it that way, I can vouch for the four now on Mars. We need to run a thorough security check on the other four, including our own."

The next G-10 board meeting pitted the will of the Americans against those who would appease those sensitive to the terrorists. However, the Americans held the trump card. They had led the way to Mars and had the people on the planet. Two G-10 members came across to form a new majority. The other board members reluctantly accepted the American demands.

None of this was made public or appeared in the news. Four people far from home had stood firm and made it clear that those on Mars would control its destiny.

The response to Mars from Dave and Gus was terse. "The G-10 board has agreed to your project plan. Your team will be responsible for planning and execution on Mars. All aforementioned resources will be provided on the six transport ships. Earth Control and JPL resources remain at your disposal."

Tim, Joan, Rob, and Mary read the message, and then stared at each other dumbfounded.

Mary asked, "What does this mean?"

Tim was sitting quietly. He was stunned. No one had expected such a complete turn in their favor. He said, "I think it means we're responsible to make things work. We will make the decisions for Mars."

Rob shook his head. "It means much more than that. G-10 and the president are giving us a free hand. What they want is results that will get them out of their fix. The problem is that we do not know what is really happening there. We need an intelligence assessment and then we need to scratch our heads a lot. Is it enough to just provide a lot of nuclear fuel in a hurry?"

Joan said, "My God. They want us to save civilization! I don't think any of us were thinking of that responsibility."

Four human beings, sitting around a table very far from home, suddenly carried the burden for the survival of humanity. Without any prompting, all bowed their heads and silently asked for help.

Some might call this experience a watershed or turning point. To a scientist, it was more of a phase transition. One might know the character of ice very well, but have no idea from the ice what water might be like. Craterwall and Mars had assumed an entirely new significance. The activities on Mars had been just about Mars. Now Mars had assumed an entirely new importance to the future of mankind—and they had suddenly become very significant. Only Gus knew what other expertise was held by the members of the Mars team.

Chapter 29:

Sabotage

ROB QUIETLY MOVED to his workstation, entered a few passwords and instructions, and returned.

Tim said, "Rob, was that related to our change in status?"

"Yes. I just activated a security firewall. All traffic must now pass through a screening process that tests for password authority and other criteria. That is also true on the Marscom. All of our incoming communications are now buffered and tested for viruses and the like. All incoming message content now goes through a new and very thorough isolated virus screen processor."

Joan said, "I thought that was already in place."

Rob said, "Yes, but nothing like this. We now have the most secure systems anywhere. Our outgoing selected traffic going to Gus and Dave is deep encrypted. They will bypass the Earth Control systems and go directly to them. No one will even be aware that we have sent them messages. There was phrasing in their last message that told me to take this action—'your team is responsible.'"

Tim said, "Rob, I think we need to bring Joan and Mary into our confidence."

Rob nodded.

"There's something you both need to know. Tim and I were called in to meet with Gus and the secretary of defense just before we launched. We met with the president and he gave us a top secret assignment. They were concerned that this G-10 arrangement might get out of control. Our assignment was to prevent international radicals from taking over this project. We must retain control over this Mars mission. A number of key phrases in our recent messages have directed us to act to assert that responsibility.

"To all outward appearances, nothing will have changed. We will be taking security measures here to ensure that nothing infiltrates our radio traffic or computer systems. We now know that we can get whatever we need through Gus."

A message was sent back in the open to those who had delegated these four people with this awesome duty.

"We will do all that is humanly possible and more to serve your needs well. We will certainly need your insights as well as your commitment in order to achieve many critical goals. We must all work together if we are to succeed."

An encrypted message also went to Gus to confirm that the security systems were in place on Marscom and at Craterwall—and that they understood their additional responsibilities.

Those involved at NASA, JPL, and on Mars became united in their determination. It was in their behavior even when they spoke casually. They were proud, focused, and part of a winning team.

An encrypted message arrived from Dave almost immediately.

Top Secret Intelligence Summary

The G-10 nations were not unanimous in accepting your leadership of the Mars Genesis Project. All of them have access to your reports and your open technology. Russia has advanced space technology of its own and has its own oil supplies. Japan has expressed interest in developing commercial operations on the Moon, but imports almost all of its oil. Most uranium metal comes from Australia. They are exposed. Japan will depend upon your enriched uranium as much as we will. Japan and Russia are now cooperative and we want to keep it that way.

Iran has enriched their uranium to weapons grade. They have developed nuclear weapons and have publicly tested them. They are using those weapons to intimidate the neighboring countries that they do not directly control. They are using the weapons to frighten the public and to foment activist demonstrations by the so-called peace movement.

The radically conservative Shiite Moslem extremists control Iran, Syria, and much of Iraq and Lebanon. They are responsible for three major terrorist strikes in the last year—two in Europe and one in the United States. All of those attacks were on modern uranium enrichment facilities. They have driven the price of crude oil high enough to intimidate the major industrial powers. Other significant oil producers are happy to benefit from the high prices. Peace activists continue to play into the hands of the terrorists.

Some of the G-10 powers are appeasing the Shiites, which is why they wanted to pack your team with their people. Those powers have shut us out of their planning. We will be changing our arrangements to include only those of the G-10 who are cooperating with us.

We have that large 2006 oil discovery in the Gulf of Mexico and we have also opened the Alaskan oil reserves to development for the duration of this crisis. That is not good for the ecosystem in Alaska or in the Gulf region, but we have no viable alternative. We are also working with all oil exploration and pumping operations with awareness of the enormous BP spill in 2010. We are aggressively pursuing oil shale, oil sands, and coal for fuels. Meanwhile, we are rapidly constructing nuclear power plants in remote areas. We—and our allies—will have an immediate need for the nuclear fuel you will be providing, particularly if Australian ores should be cut off.

The Mars uranium ores are much richer than those on Earth, but they require just as much enrichment to become nuclear power plant fuel. You will be enriching only to 5.0 percent U-235. Weapons grade is 90 percent U-235 so you will not be involved in weapons production in any way. The main reason for you to enrich uranium is to fuel your own power plants and reduce the mass of the enriched uranium you are sending to Earth fivefold—and we anticipate that you will be expanding your base and its operations rather aggressively.

Once you are able to enrich uranium and send rockets back to Earth, you will be seen in another light. You will be providing fuel to replace oil and you will be seen as a power unto yourselves. Please be very careful with how you express yourselves in any communication you send back. Also be very careful to watch your backs. You must double-check everything you receive for AI robots. Discovering uranium on Mars could be a mixed blessing.

Most people see your successes as nothing short of miracles. You are extremely important as a resource, but we also need you for the image you present. If you can make it on Mars, then we can hold things together here. You represent the positive spirit that we must revive and sustain.

David Bagnal
NASA Administrator

Mary shook her head and said, "It looks like we just jumped from the frying pan into the fire. Can we trust the four new people?"

Rob said, "You can bet that Dave and the others are taking a very close look at their backgrounds right now. They'll let us know if someone needs to be changed. They'll also be assigning a security detail to check and double-check what's being shipped to us." He smiled grimly. "I worked in military intelligence regarding computer security. I also put in some time with the NSA. That does not show in my records. I am now our security chief."

The two women looked at each other.

Joan said, "Is there anything else you can tell us that will make us feel better?"

Rob said, "If I had anything like that, it would not be something I could tell you." He offered an encouraging smile. "I can tell you this. I have already changed our procedures on incoming messages and software. For example, all messages will be screened for any viruses in a separate and isolated processor. All software will be thoroughly tested in a suitable piece of equipment set aside just for that purpose."

Tim, looking for a lighter mood, said, "We do have those talents we revealed a few months back."

"Our incoming *Pioneer* ships are carrying equipment and materials that will make it possible for us to implement those discoveries. Let me give you an example. We have frozen human eggs and sperm—including our own—that we brought so we could fertilize in vitro then transfer the eggs back into Joan or me if the need arose. We can actually go a lot further. After fertilizing the egg, we can grow it in the laboratory and transfer it to an artificial uterus that has all the functionality needed to promote its growth as an embryo and into a normal baby during the period of normal gestation. We could greatly expand our human population as we see fit without having to wait out the individual normal gestation of each pregnant woman.

"This process still uses normal unmodified eggs and sperm from humans and simply allows gestation outside of a human body. We would not be creating anything but a normal human being. This would also make it possible to rapidly increase the human population here and avoid the great expense and risk of transporting more people to Mars. It would require us to care for and educate a fairly large number of children."

Joan said, "We're speaking of a watershed change in direction for our undertaking."

Tim frowned and said, "I suggest we make inference in our next message that we would be using all our knowledge and resources."

Rob was in agreement. "We certainly have our hands full for now."

*   *   *

Their communications took a hit. Their normal systems sensed the activity and automatically made them dormant. Rob saw an analysis of the cause of the hit, including the nature of the incoming message's character. He immediately sent that back to Gus as a secure message. The source of the errant message was Mexico. A combat team reached the spot within hours, but—other than a large antenna—they found nothing. The evidence showed that it was manufactured in China.

Communications systems were activated once the Mexico mission was completed. There was another hit on the communications systems the next afternoon, this time originating from Earth Control in Houston. The programming was embedded in a message to a robot on Mars.

This time, though, Gus received a text message on his cell phone from his son, high school student, asking him to contact a rogue cell phone number. He immediately brought in his own security people who listened in on his call.

"Dr. Hoover, please excuse this interruption. I am Dr. Ahmed, working with the robotic direction project on Mars. My family and I are threatened for our lives by terrorists from our home country. My son goes to school with Gus, Jr. He arranged this contact. Please help us."

Gus said, "We know you're in your normal work area. Where are you precisely?"

"I'm in the restroom. A custodian, Mohammed Arawat, is the one who stood beside me with a gun while I sent the message. You must capture him quickly. He has two other accomplices who may be near my family."

"Your work building was locked down immediately when you hit the send key. We have security going to all of the phone locations of your family members right now."

"Mohammed Arawat was profiled as were all of his contacts. All of their vehicles were being tracked.

"The security team moved in on your house as soon as a contact vehicle was seen approaching it. They caught the two accomplices as they approached the entrance to the house. They're in custody. The man who threatened you is also in custody. We're picking up all of your children and taking them to a protected location."

Two days later, a very encouraging message came from Gus. Josie, Arnie, Ho, and Norika had all been approved to augment the Mars Team. Tim sent an immediate reply through Gus to welcome them.

Ho, Norika, Josie and Arnie

You are all far more than welcome to our Mars family. You will find that we are very casual and that we are more a family than a team. Those of us here on Mars have come to instinctively know how our friends will respond in any given situation. You will also find that, while our robots are servants, we interact with them very informally, even talking with them as people, using familiar names.

You may want to bring along personal items to make you more comfortable so far away from your families and home. This is your new home.

You will have very little time to prepare yourselves for your departure. You'll be learning a lot about space travel and what it will be like here. You should not worry about our activities here as we will help you adjust much more quickly once you arrive.

While your technical expertise will be critical to our success, you as individuals are exceedingly important to our relationships, and your becoming part of our family is at least as important. We are not clannish and we expect to be very open with you from now on. Welcome.

The pioneers were soon bonding. Ho and Norika had already established a relationship as experienced NASA professionals. Josie and Arnie came in from the outside, but quickly found their new partners more than interesting. They needed to acclimatize to the NASA family.

Gus brought them in to his office. He said, "Thank you for coming aboard. You're all very important to the future of people on Mars and much more.

Arnie, Ho, and Norika, you will be focused on getting the Mars economy jumping with abundant nuclear power. Finding rich uranium was a godsend. It opens the door to much more rapid growth than we had imagined.

"Josie, you'll be working with the army of robots and robotic equipment and their AI technology. You complement those already on Mars. You will be bringing equipment that you can use to produce robots and robotic equipment and that will give us the physical means to grow. We can't produce more people, but we should be able to produce hundreds of robots. With those resources, we can multiply our industrial base very fast. You need to keep track of all those resources up to the second.

"You were all selected as much for your personalities as for your technical astuteness. The current Mars Team is as close as brothers and sisters—and I mean that in the most positive way. You can talk with them on any subject and nothing is too personal. Our greatest concern in doubling the size of the team is maintaining the closeness and camaraderie of that group. You need to fit together by thinking alike on strategic matters.

"Josie and Arnie, you are not familiar with our way of doing things at NASA and JPL. Please just jump in. You're already inside. No tests. We will just be helping you get ready."

Josie was a dynamic redhead and remarkably insightful in applied artificial intelligence. Arnie was a sturdy uranium mining engineer with plenty of Aussie spunk. Ho was a meticulous young man with many innovations to his credit in designing and building carbon fiber structures. Norika was a small bundle of energy with degrees and practical experience in enriching nuclear fuel.

A special link was established between the Mars Team and the Pioneer Four. Every effort was made to include them in the Mars perspective and to bring them close with each other. The Pioneer Four were included in many of the day-to-day happenings within Craterwall—even from well before the launch.

The Earth Control team was still their twenty-four-seven project team that handled all of the routine construction within Craterwall. The Pioneer Four were given practical experience with the Mars robots while the controllers guided their way. Since Mars was deep in the middle of its winter, all outside work was shut down. All activity was synchronized with the Mars work clock. The new team members worked on the Mars schedule.

Each member of the Mars Team was assigned a partner with the incoming group to make them welcome and quickly bring them up to speed. The partners dove in with all of their energy. Rob worked with Arnie, Joan worked with Norika, Mary worked with Josie, and Tim worked with Ho. They made it a point to connect with each other at the end of each Mars work shift to talk over what was going on and why. They always got together for an evening meal after their discussions with the Mars foursome. This way, they transferred their partners' experiences to all members of their own group.

There had been many evolutionary changes in the lifestyle of the original team—beyond having the children to care for. They had jogging routes around the corridors that they faithfully used twice each day.

Their diet had moved toward vegetarian with emphasis on foods that would keep their digestive systems moving in the low gravity. Sports such as badminton compensated for their low gravity. Volleyball was also useful as well as putt-putt golf. They even found a way to compete with their Segways. The robots became very good at refereeing.

One of their rooms had been designed to contain a circular swimming pool with enough depth to allow for diving. Splashed water had a way of hanging in the air and care was made to avoiding a mouthful of suspended liquid.

The farm had exploded in population. Everyone had a favorite, particularly with the ferrets. The newcomers would be bringing more animal and plant diversity. The animals were stimulated into athletic activities of their own by the robots.

CHAPTER 30:

REINFORCEMENTS

LAUNCHES WERE ALWAYS exciting. The first two ships carried only cargo—although it was very precious. Both were launched without undue delays. Now that Mars was related to rockets returning to Earth there was a group of activists that saw the project in a negative light. There was even a terrorist attempt to destroy the ships as they sat on the launch pad.

The third and fourth ships carried the pioneers. Ho and Norika were on the third ship. Josie and Arnie were on the fourth ship. *Pioneer 3* had local demonstrators at the gate, but local guards were able to control their anger with only a few arrests.

With *Pioneer 4,* it was two groups, the demonstrators, and a well-armed group of terrorists. The latter made a concerted attempt to break into Cape Canaveral to damage the fourth ship. Army troops were placed near the launch complex. The violence developed into a pitched battle. Demonstrators were caught between the troops and the attackers. Three demonstrators were killed and many were injured. Ten attackers were killed and the remainder captured. The attackers were clearly mercenaries from Middle Eastern countries.

The fifth and sixth ships did not carry people, but they did carry critical equipment. The demonstrators were scared off by their experience with the aggressor attack against the fourth ship, but there was still one attempt by Shiites to fire an armed rocket toward the last ship. They were captured before the rocket could be launched. Craterwall was clearly not seen as housing an innocent scientific expedition any more. The growing Mars Team was in the thick of things.

The ships were strung out seven days apart across space. Their experiences with the previous interplanetary transits had taught them where to augment their redundancies. However, nothing could provide protection from the infrequent stray meteorites.

Gus, Dave, and the four colonists already on Mars followed progress attentively. But their greatest concern was with the prospect of a very large global Mars dust storm. Landings were expected to take place regardless of atmospheric conditions. They could not wait out an extended storm. Powerful radio transmitters were positioned in a grid distributed that covered the elliptical region of anticipated landing sites. When an approaching ship descended into the dust cloud, it would lose guidance from the GPS satellite. As it approached the surface, one ground radio transmitter/receiver or another would make contact and track the ship down to the ground. The radio link would also give the ship its exact position and track it until landing. The information would allow the recovery vehicles to reach the ship almost immediately.

Another precaution was made. Twelve safe haven shelters with critical supplies were located along radials from the central landing site's tarmac, extending between twenty and thirty kilometers distance. They provided high-energy radio to penetrate the dust cloud with pulse pings. The shelters were dug in to provide shelter from any errant solar storm emissions that could easily kill anyone exposed. Supplies included oxygen and other gases such as methane fuel and foodstuffs. There was also a fiber optic communications link to Mars Control.

The Mars Team could not lose sight of their immediate mission with new people coming and the aggressive radicals on Earth. In that regard, they had to work from the assumption that all those joining them were of their mind. One way to do that was to get them into conversations on the threats faced back on Earth—on the pretext that they had been out of contact and needed to hear from those who had lived it firsthand. That would wait until they arrived. Body language could mean a lot.

In the meantime, the Mars Team had to face their first full winter burrowed beneath a comforting shield of ancient meteor impact spoil. The onset of winter in the northern hemisphere meant that the southern hemisphere was turned toward sunlight while the pronounced elliptical orbit of Mars drew the planet closer to the sun. The dry ice cap on the southern pole had become thicker than normal during the long winter.

As that pole warmed, the evaporating polar cap again turned dry ice into enormous volumes of carbon dioxide gas that flowed toward the equator. NASA forecast a heavy season of storms, perhaps even of global expanse, stretching well into the northern hemisphere as the excess carbon dioxide streamed toward the North Pole where it would freeze into solid dry ice again.

Human-initiated activity outside was quiet once hibernation of the outside facilities began. Inside, the pace only quickened. The corridors and rooms along the extended Inner and Outer Ring Corridors and the new Southwest Radial Corridor (SWRC) were bare and not yet Marscrete sealed.

The atmosphere along the SWRC was Mars native, pure and simple. It was also just as cold as the arctic exterior—except for the extension of the Outer Corridor. Its compressed Mars atmosphere was heated from the nearby nuclear furnace so that they could periodically warm and occasionally repair the equipment. Since there was ample electricity and heat inside the older structure, there was no further drain for outside activities.

A stockpile of mixed Marscrete powder for plastering the interior walls was stored just outside the SWRC entrance, which was not closed. Utility lines were being run throughout the new space, connecting to sections that had been embedded in the Marscrete block walls.

*   *   *

The rockets were well on the way to Mars. The ships were more reliable and more comfortable. However, they faced a new challenge.

The dust storm out of the Arctic the prior spring had been small compared to the torrent of storms from the Antarctic during this southern spring. The melt at the South Pole was earlier and more vigorous than normal. Three relatively small storms were developing with some intensity. The forecasters expected them to grow and combine, spreading widely over the planet. They did not know if they would reach as far north as Craterwall or if they could even combine and become a global storm. If global, the northern cap could grow larger than normal and the cycle could continue.

Should the storms extend northerly, it would prolong the northern winter by obscuring the sun in the north and making the winter even more intense.

*   *   *

The interior construction work was pushed rapidly since their lives depended upon it. Air locks blocked the intersections within the ring corridors and along the Southwest Radial Corridor. Sealing the walls of the corridors and rooms with Marscrete paste began from the Inner Corridor air lock at the West Radial Corridor and was extending to the Southwest Radial Corridor, outward along that corridor to the junction with the Outer Corridor, then from the airlock near the Power Wing back to the Southwest Radial Corridor.

The idea was to keep the construction equipment backing out from the interior toward the Southwest Radial exit onto the surface. The exterior atmosphere within the sealed space was changed to heated air that they could breathe and standard interior atmosphere.

They opened the Inner Ring Corridor and the attached rooms for use very gradually. It was all going to be used for farming and human habitat space. Plants and animals were moved over to the new farm space and the human habitat space across the West Radial Corridor was expanded for the incoming human occupancy. The idea was to place all of the humans in adjacent space north of the West Radial Corridor.

The original Mars Team wanted to make their new companions perfectly at home and intimately involved in their daily activities. They must feel just as involved and actively participating as the original four.

Meanwhile, the outside temperatures on Mars plummeted and the deep freeze of the northern winter was upon them. The nuclear fuel rods were inserted deeper within the furnace to increase the heat being generated. The interior temperature remained optimal for the living things.

Mary was intensely busy, occupied with all of the new interior space. Rob was often called upon by Mary to provide expanded habitat facilities: frames, tanks, and troughs to feed the livestock, water troughs for tilapia and catfish, hydroponic troughs for nutrient-carrying water, circulation systems, nests for chickens and rabbits, roosts for chickens, cages, and more.

Mary and Jeb were continually moving soil-rooted plants, hydroponics, and animal birthing and growing facilities. Mary realized that the weights the humans carried were just as important for the animals—and they also needed an exercise regimen.

Food was harvested and stored in tightly sealed containers—and then frozen. There was no assumption that the living things would continue to flourish. They were stockpiling for future needs.

Joan and Tim were the future. They were preparing for industrial expansion along the Outer Ring and were in frequent contact with the incoming colonists, JPL, and Earth Control.

The first step was to construct structures of the needed diameter and to test them in extremely fast configurations to see if they would hold up. The main reason for using carbon fiber laminates was to reduce the weight of the cylinder walls while retaining tensile strength in the walls.

Tim and Joan were also working with JPL toward constructing a full, commercial-scale nuclear power plant. That was to be constructed in parts that would be assembled once the parts arrived and the weather warmed.

The decision had been made to keep high tech progress secret. That changed everything on Mars and in their communications with Earth. In truth, they were making exceptional progress in all quarters, but their reports were toned down to sound somewhat pessimistic.

Rooms branched off of the Southwest Radial Corridor and along the extensions of the Inner and Outer Ring Corridors. All were completed and occupied.

The technical projects were also on schedule and tested better than their design criteria. A substantial amount of uranium ore from the fossa had been refined to pure uranium.

A tunnel had been bored a good kilometer outward from Craterwall to a smaller nearby crater, but they ran into a problem when the tunnel collapsed and three robots were destroyed. Another tunnel was constructed, but it was carefully reinforced with Marscrete blocks.

The uranium was transformed into uranium hexafluoride gas and tested for enrichment in the first centrifuge stage. Incessant tests were made to withdraw the enriched uranium, but the desired enrichment level was not attained. Considerable communication with the American Centrifuge project discovered a seemingly trivial oversight and finally was successful.

The full centrifuge array would have eight units. The centrifuge had been set up underground. The first unit successfully raised the concentration of U-235 from 0.7 percent to 1.4%

As the first landing approached, the last of the Mars manufactured parts for the new full-scale pebble reactor nuclear power plant were undergoing final tests. The Mars alloys were not up to par, but repeated adjustments to the process finally gave them success. Rob and Joan were being driven to exasperation. The extended near-global dust storm was wearing on all of them.

Meanwhile, the master plan for Craterwall, including all outlying activities, had matured considerably. All transport had previously been by fueled vehicles. The new concept was to build a rail system that would allow rapid enclosed vehicular movement between all of the facilities.

The original primary objective of going to Mars had been extensive exploration of the surface. They were now becoming eager to commence that activity in earnest. Over the next Mars year, they would be reaching out in all directions. Their first goal was to reach and explore the long Kassei Valles that reached deep into the Tharsis Bulge and much closer to the Mars equator. It would carry them near another cluster of fossae and some extraordinary landscapes. The Earth-Mars planetary opposition occurred earlier in the Mars year with each succeeding conjunction because the Mars year was just over two Earth years in length. The ships would arrive before the spring equinox when the weather on Mars would be colder. They would arrive during the northern hemisphere dust storm season.

Two months before the arrival of *Pioneer 1*, the weather was still bitterly cold, but the atmosphere was clear of dust—despite the forecast of an intense, extended dust storm season.

The team was continuing its construction within Craterwall, but the seemingly endless confinement within their habitat was wearing. Before winter, they had worked outside with robots and robotics—and they had infrequently made outside excursions. For four months, they might as well have been confined to Earth.

They had spent four years almost entirely immersed in the Mars project—followed by six months in space and sixteen months of intense work—constructing the underground living and working space. And while the children were a welcome addition, caring for them was becoming another burden among many, many others. They were still facing two more months before the new ships arrived—with the prospect of landing them in what promised to be a lengthy, intense global dust storm.

They carried the additional burden of being the inspiration and promise for the people and nations of Earth. They desperately needed to bring the six new ships down safely—and perhaps, most of all, they needed the support of four more humans in their daily lives.

Baby John was eleven months old and baby Elizabeth was ten months old. John was trying to walk and Elizabeth was actively crawling around in their shared playpen.

*   *   *

Mary returned from her rounds with the animals. Tim and Rob had just returned from a run and were returning to the workroom. Joan was at her workstation. Martha was fixing lunch. Crusoe was straightening up after the children.

Most of the messages from Earth were directed at the robots and robotic equipment by Earth controllers as they continued the seemingly endless expansion of the Craterwall complex. They were logged, but not really tracked by the Mars team.

The latest message was different. It was a Mars weather forecast from Earth Control introduced by Gus.

> I've reviewed this forecast in particular because it introduces the first sign of carbon dioxide sublimation from the North Polar region. The polar cap is the most massive we have seen since we began monitoring these accumulations. This is ominous in that it will fuel the dust storms you'll face. We expect you'll

be socked in by dust within two to three weeks and this could last throughout the spring season.

You've constructed a high-powered parabolic radio transmitter, which we hope will be able to punch through the dust. The incoming ships will also have stronger transmitters aimed at your site. Still, you will not be able to work outside during most of this time because of the wear it would cause on your equipment.

Our fleet of *Pioneer* ships is faring well in transit. You'll have the latest on their condition before the dust settles. You'll have their exact arrival times. Still you will have to dare to meet each of the ships, rescue the crews, and retrieve their cargo immediately upon their arrival.

There was a second encrypted message.

Conditions here remain tense. Terrorist action is becoming pervasive around the world. Nuclear power plants are being constructed throughout most of Europe and North America as well as Japan and Australia. Hybrid and completely electrical automobiles are quickly becoming the most common. We have stopped encouraging ethanol production as just too costly and punishing to the agricultural economy.

Please keep sending your reports of positive achievements addressed to the public. The people need that encouragement in this time of trial.

They read the message and shrugged it off. There was little new except the indication of an early beginning to the dust storm generation out of the North Polar cap.

Tim composed a message to *Pioneer 3* and *Pioneer 4*.

Ho, Norika, Josie, and Arnie,

We expect you have received the latest weather report from Earth Control. We're very experienced now in handling incoming landers—even in severe weather.

Our facilities are improving. Your home space is nearly complete, although we lack the equipment you're bringing. Our systems for managing the atmospheric environment are keeping compositional gases well balanced despite the increased plant and animal populations. We now have a dozen ferrets roaming our living space and the finches are nesting in the branches of most of our bushes and trees.

Our livestock population is multiplying as well. We have to control the rabbit population and the pigs are enough to supply our needs when you arrive. We're shearing sheep for woolen fabric as well as harvesting cotton and flax.

John and Elizabeth are growing normally. John is toddling about a bit more than we would like.

CHAPTER 31:

SECURE PRESENCE

FOUR WEEKS BEFORE the arrival of *Pioneer 1*, the dust was thickening by the day. Visibility had dropped from a kilometer to hundreds of meters to perhaps ten meters or so. It was again the same as looking though a dense fog.

The last transmission to Earth Control had been a week earlier. There was nothing to breathe outside and the visibility was zero, but everything was humming inside. Everything was operational. They were a pocket of life on a far away world and they were optimistic. They had been selected for their determination to survive and would not be deterred.

Their inspiration was their children and the everyday reminder from the animals that life goes on. Lambs, kid goats, newborn rabbits, and baby chicks did not know that they were a long way from their parents' homes and had no idea that Mars was a very foreign place to live.

Mary walked into the work area carrying a freshly born kid and Jeb followed her with another. Its nanny goat mother tagged along behind concerned that her offspring might be lost.

"Hi all. I want to share our latest additions. Flopsie here has given us twins. This is very special. I think they're as cute as can be."

Everyone stopped their activity to gather around Flopsie and her lambs. They each inspected the little ones and gave Flopsie some scratches and pats. A significant new birth was always a reason for celebration—a reminder that they were the source of such miracles.

Tim said, "You know, we work twelve hours most every day without doing much to see the beauty of what we have created. Every one of us does something special every day and we take those sparkles of light for granted. Let's call a break and give ourselves a happy time—a party."

And so it was.

*   *   *

Each week they sent a communications rocket high above the dust layer, each one carrying a message to Gus and their incoming colleagues and picking up a waiting message from the Marscom satellite. Two days before the first expected landing, the message from Gus was encouraging. All of the Pioneer ships were doing well as was their cargo. Tim included encouragement to the newcomers and reminded them that the Mars Team was ready and waiting.

Pioneer 1 carried precious equipment and diverse livestock and plant life, but no people. They would use this landing to test all the new technology in the transport ships.

They were up early the morning of the expected landing. A perfectly positioned landing would put the ship in the middle of the new space port. No one expected perfection, but there was a probable landing ellipse. Tim, as before, was at the port with *Venture 1* warmed up and ready to redirect an errant lander. Joan, Rob, and the robots sat in a nearby hauler poised to scramble to the site of the actual landing. They were all in pressure suits and the communications had been checked and rechecked.

The scheduled moment of the landing arrived. Mary fired off the latest communications rocket and received back assurance that *Pioneer 1* had executed the entry and descent. All indications were positive. All that remained was the actual landing—or so it seemed.

The grid of high-energy radios surrounding the tarmac began sending out pings that would awaken a response from the landing craft. Tim scanned with all the useful instruments at his disposal, but there was no response from *Pioneer 1*. The twelve outlying radios at the emergency support sites likewise pinged with similar results.

Tim said, "We have a lost bird! At this location, we have no ground equipment to help us find it."

If *Pioneer 1* had landed safely, then the onboard robots should follow a procedure that would help them find their way to a safe haven emergency support site, but that could take quite a while. Tim, Rob, and Joan returned to Craterwall and took up a vigil at Mars Control.

Time crept. The team was facing a challenge that offered little hope or resolving in time to save the living things that the robots should be hauling. It was cut off from all communication by the dust storm. Everything depended on how well the robots and equipment executed their instructions and how intelligent their AI systems may be. The landing was to have been executed at 09:23 hours.

The tension increased by the minute. A small mistake in the location of the landing should have been resolved fairly quickly, but the terrain to the southwest was fairly rugged and it could be difficult for the AI to navigate that kind of maze for any distance.

The anxiety was evident in their fidgeting. Tim could do little except wait. Joan camped out near him. The others kept watching the time, but avoided interfering with Tim's checking and rechecking. A hauler should be coming into range of a radio beacon. The first hour had been hopeful. By the fourth hour, the team had become resigned to what seemed inevitable—all of the plants and animals would perish.

At 14:45 a beacon received a ping. It responded with directional information to the hauler. Twelve minutes later, the lost hauler pulled into the shelter and the hauler's robots linked their com unit through the shelter unit to Mars Control. The ship had safely landed thirty-four kilometers southeast of the intended location.

The relief was palpable—even though the rescue of the remaining hauler was still unrealized. The robots on the *Pioneer* hauler immediately replenished its fuel and atmospheric gases. Tim dispatched two haulers with robots and supplies from Craterwall to the safe haven. They reached the safe haven by 15:53 using a rough roadway. A *Pioneer* robot would serve as a guide to Craterwall. It arrived at the west radial entry lock at 17:13 hours and was brought inside for off loading.

Two Craterwall haulers back-tracked the errant hauler in the morning. However, even with tracking electronics, the search for the landing site was time consuming without GPS. After loading equipment and supplies and calling the second *Pioneer* hauler, the small convoy began their return journey. With a brief stopover at the safe haven, they completed the journey to Craterwall before dark. It was by far the most dramatic demonstration of AI adaptability so far.

Tim had immediately begun an after-action review and deciphered the data. The landing site coordinates were changed from those originally recorded in the computer memory, but the next questions were why and how.

The original coordinates were still in the lander's permanent computer memory, but another set of coordinates had been hard-coded and took priority. Had that been intentional sabotage or the result of an error in programming? It was a problem for Gus to resolve.

Three days later, the information was transmitted through the weekly communications rocket to Gus and JPL. Unknown to the Mars Team, a corrected entry-descent-landing program was then sent by JPL to *Pioneer*

2 and the other incoming ships. Gus initiated an investigation of the hard-coded data, but all indications were of a programming mistake to facilitate early testing.

Everything was ready when *Pioneer 2* approached. The landing went exactly as planned, putting the lander on one corner of the tarmac. The relief for the Mars Team was immense. Both loads of plants and animals arrived in good condition. They were all placed in a treatment program to bring them around. Two small Jersey cows were of special interest. They would be inseminated as soon as they regained strength. A small herd of cattle was needed to provide milk for infants and growing children who needed the calcium.

Sorting out the plants, animals, supplies, and equipment was handled automatically by the robots. They had manifests and assigned locations. Mary and Jeb managed the arrangements for the living things.

They were relieved and apprehensive. The surprises were wearing. Tim called a break and a day of relaxation that was very much needed.

The focus of planning and activities immediately shifted to the next two ships and their human passengers. Ho and Norika would arrive on *Pioneer 3* in less than a week. The regular communications rocket carried good news to Gus and the others.

The usual arrangements were made for the landing. The ship executed its entry and descent as programmed. Tim was on the remote console as a precaution against a faulty landing.

Tim monitored every move as the lander slowed and entered a vertical descent. That was when he sensed—more than observed—a deviation. He switched to remote control so that he could take over the remaining descent, but in that brief instant, the ship began a radical move away from its planned path.

Tim's training kicked in. He sent corrective maneuvers to the landing craft, bringing it back to a vertical path. The ship was away from its targeted landing spot, but still in an area without large boulders. He brought the ship directly down to the ground, not wanting to chance further errant behavior. Immediately upon landing, he killed all of the rockets and turned off the master switch on the landing computer. He wanted this ship to stay put.

"Rob, get them out of that ship right away! Something manic was taking control. I was lucky to get it on the ground."

Rob, Josie, and the robots that accompanied them in the haulers were already climbing up the ladders to the cabin and urging Ho and Norika to execute an emergency evacuation. The people in their pressure suits were shoved into the cabs of the haulers. Meanwhile, the cargo lift was released and the robots were moving all of the equipment into the hauler beds. Rob and Josie pulled the haulers back a safe distance just as the lander exploded.

Meanwhile, Tim had climbed aboard his hauler and had arrived on the scene.

Josie was clearly shaken up. Rob was angry as hell and cursing. Tim could barely control his temper. He went immediately to Ho and Norika in their haulers. They were stupefied.

Tim spoke with obvious relief in his voice. He said, "Thank God we got you out. Welcome, but we would have liked to avoid the fireworks."

From Mars Control, Mary asked "What happened? Are you okay?"

Rob said, "Mary, the ship was set to crash then explode. Miraculously, we are all fine and we even got the stuff out before the explosion. Ho and Norika are inside the haulers."

Tim turned to the others and said, "This is not the end. If someone sabotaged that ship, then *Pioneer 4* is also set for trouble. Josie and Arnie are riding inside of a bomb. We must message this to Gus as soon as we do a little sleuthing. Those SOBs were not just aiming at Ho and Norika—that blast was also intended to take you out!"

On the ride back to Craterwall, the team was fuming. Mary was waiting for them. There was a determined reception when they got the newcomers out of their pressure suits. Everyone hugged vigorously as if by doing so they might make things okay.

The robots took care of the animals, plants, and cargo. Ho and Norika were carried to the habitat. Crusoe took over to make them comfortable. Martha brought them some of her best soup and helped them eat. Mary and Joan stayed with their new friends. Tim and Rob went into deep conversation off to the side.

Rob looked deeply into Tim's eyes and said, "What happened with the landing?"

"I'm not entirely certain. I was focused intently on the monitor when something clicked and I took over for a remote landing, overriding the automatics in the ship. I don't even remember my specific actions. When the ship was stable, my first conscious act was to expedite the landing. Something was telling me that I had to get the ship on the ground right away. When it reached the ground, I immediately popped the hatches and lowered the cargo pod, killed all the power sources, and called to you for help. You know all the rest."

"Was there something that told you to get them out right away?"

"Just training. It's standard procedure in military aviation to immediately get people away from a crash site. You have to expect that there could be an explosion."

"What do you think happened? We have to send a report to Gus. He'll need all the time we can give him to figure out what to do for *Pioneer 4*."

"The program that controlled the landing was obviously tampered with. Furthermore, there was an independent firing device for the explosives that was timed to go off after a delay. The ship actually had a perfectly soft landing. The explosion was not activated by a rough landing—although there could have been something to do with the landing pads."

Rob shook his head. There could be no answer.

Their discussion continued for another half hour and then they composed a message to Gus.

The unexpected message arrived at Earth Control, encrypted, and addressed for Gus only.

Highest Priority

Pioneer 3 was destroyed in an explosion after landing. The landing software went haywire and attempted to crash the ship, but Tim managed to get it safely on the ground. All personnel and cargo were very rapidly unloaded before the explosion. The delay was obviously intended to allow time for us to approach and be killed in the explosion.

You must solve the riddle of the software malfunction and fix that problem in time to protect Josephine and Arnold.

We will operate on the assumption that you have failed and do our best to protect our new friends.

The normal message this week will be false, indicating that the ship crashed with fatalities. That is to avoid alerting the culprits while you seek them out.

This message is only for you unless you wish otherwise.

Mars Team

Gus dropped his head onto his hands. Somehow the Mars Team had produced another miracle, but now he faced an almost impossible prospect.

Gus was fuming. He immediately contacted the team leader at JPL after sending him a copy of the message.

"You have a mole in your organization! Our people on Mars saved themselves—and us—from a catastrophe. If their plan had worked, we would have only Mary on Mars now! I am not happy!"

The JPL director was dumbfounded. He said, "I hear you. I can't believe we have anyone on our staff that could do this. It would be two acts of sabotage—one to corrupt the landing programming and another to attach a bomb that could sense when the ship got on the ground."

"Well, get on with it *now*. I'll have the FBI and CIA alerted to this crime and you go through your program code immediately. Do you realize what had to be behind this?"

Everyone from the top down was thrown into immediate action. The JPL team sent a query to the central computer on *Pioneer 4* for the date of the compiled program. It was more recent than the JPL-compiled copy. Someone had evidently modified the source code and inserted the spurious content.

They immediately recompiled the source code to get a clean copy with the current date. It was transmitted to all of the remaining in-transit ships. Because it had a later date, it would automatically replace the older copies.

That did not solve the problem with the explosives. They had no way to detect and disable an explosives trigger that was completely independent of the ships' systems. The FBI and CIA both investigated the security problem at JPL. It took forty-eight hours of extreme interrogation to find the terrorist culprit, but that did nothing to solve the problem. There was still no way to deactivate the explosives.

*   *   *

The Mars Team sent the regular message on schedule with the false information. Gus intercepted the message and released word to the press that the landing had failed with all aboard killed.

Gus had also sent an encrypted message to Mars, bringing them up to date on the investigation. He also let them know that the explosives were still live. He added, "You must give priority to rescuing Josie and Arnie. Do not risk people in unloading the ship."

Tim had already sent word to *Pioneer 4* that they would be handling the evacuation expeditiously. Josie and Arnie were very tense, but after the rescue of Ho and Norika, they had real hope that they could survive the landing.

When the ship came in for landing, the Mars Team knew the landing program had been corrected so they let the ship land on its own. Tim took control immediately prior to landing. Upon landing, he released the cargo pod and the personnel hatches immediately and then cut power to all systems.

Josie and Arnie made an emergency exit into Rob and Joan's arms; with robot assistance, they all sprinted away from the ship. The robots grabbed all cargo possible, moving it far enough to be out of harm's way. They were finishing when the ship exploded, knocking the robots flat.

Tim was just reaching the scene when the ship exploded. He joined the others at one of the haulers. "How are you doing? Do you need any help or treatment?"

Arnie said, "I thought we Australians gave an enthusiastic welcome to new arrivals, but this is ridiculous."

The robots that had been blown over assured them that they had performed an electronic and physical self-test and that all was okay. Everyone at the landing site made an immediate trip to Craterwall where they were greeted by the others with enthusiasm and settled in at the habitat.

Mary was still anxious. She said, "Tim, do you think this is the end of it? Will the other ships be set with explosives?"

Tim's said, "I expect not, but we won't take any chances. The landing software has been corrected and they most likely would not take chances with being caught setting the explosives. They would expect everything else would take out the ships."

Indeed, the unmanned landings were by the book with no explosions. The robots had quickly unloaded all of their cargo with no surprises.

*   *   *

Meanwhile, the newcomers were adapting to life on Mars. The expanded interior of the Craterwall complex was useful and fully charged with habitat atmosphere. The power plant was operating at near capacity with no room for including further power consumers—no construction could be made without trimming other power consumption.

They were capped out, but they had the equipment to construct four nuclear pebble reactors and the basic equipment to construct a commercial-size fuel rod reactor. They also had the latest robot control software and enough parts to upgrade all of the other robots. They had enough critical parts to grow their army of robots to almost five hundred.

*   *   *

The dust storm showed no signs of abating. The periodic weather updates from Earth were not encouraging. The entire planet was covered in this storm. The prospect was for eight more weeks before any chance of relief.

Joan and Arnold were busy reviewing plans for aggressively expanding uranium mining and refining. Tim and Josie were setting up the new robotic equipment, upgrading the software for the older equipment, upgrading computers, and maintaining the robots. Ho and Norika took over a room where they were setting up and testing components of the first full-size centrifuge. They also began practicing and testing carbon filament and epoxy construction of a full-size centrifuge casing.

The newcomers found time for themselves. They were managing to get around and explore the facilities while recovering and gaining strength.

They were enjoying a short afternoon break when Arnold expressed a view he had been holding to himself. He said, "You know, our veterans are a bit more than I expected. They are very self assured and almost communicate telepathically."

Josie said, "We knew them before they left for Mars. They were very capable even then, but now they exude confidence. Before they were good team players, but now they're like Super Bowl champions. I really like this. Their confidence is contagious. I feel uplifted even with our explosive landings. This is really good!"

Ho smiled at Norika. He said, "I see something special that comes from overcoming extreme challenges. I'm glad we're on the same team. I expect in time we too will feel that way. For now, I just want to learn from them. How do you like wearing the gold weights?"

Norika said, "The golden look is exotic and novel, but really just added weight to compensate for our lower gravity. When our ship exploded, I was really wondering how things could work out well. Now I know it will. We are cohesive. Even the robots fit well in the scheme of daily activity."

*   *   *

Rob was the maintenance supervisor, monitoring the entire facility and overseeing all of the operational production processes. Mary had her hands full working with Jeb and his helpers in managing the many new species. Martha and Crusoe continued in their household duties, freezing and preserving foods, and minding the children.

Everything was recycled, including composting of spent vegetation, using animal carcasses and animal waste for soil enrichment, and irrigating vegetation with white water from the dish washing and bathing. Other waters were distilled and placed back into the distribution system. No waste, discarded containers, or paper were trashed.

The newcomers would be ten weeks or more in regaining their physical strength and adapting to the new environment. Their small pool was a welcome diversion.

The facility now included the Northwest, West and Southwest Radials and the joining Inner and Outer Ring segments. One-quarter of the crater's wall was now constructed as living and working space. Domed rooms lined all of these passageways. The Segways were used for rapid movement; they mostly used the passageways as a jogging track.

*   *   *

Joan and Mary were nearing term on their second pregnancies and Josie and Norika were intent upon becoming pregnant as soon as their physical condition allowed it.

With the continuing dust storm, the robots were being directed to sensitive projects during waking hours and assigned routine tasks during human sleep hours. Earth Control would take over more detailed direction once the dust abated.

CHAPTER 32:

ERUPTION

THE ENLARGED MARS Team developed synergy over the next twelve weeks. The original four had hungered far more for companionship than any of them had realized. The interpersonal activity was so appreciated that they wasted no time in finding ways to interact. Low-gravity sports were played and even sports such as bowling in the blind extensions of the circular passageways were organized.

The Outer Ring construction was pushed across the South Radial intersection and on to the Southeast Radial. That allowed robots to survey the pebble reactor pit site and to begin pit excavation leading to lining it with Marscrete blocks.

The components of the reactor and its power generator were installed immediately. Power lines were strung back across the south intersection and on to power line connection with the live terminus at the Southwest Radial. The generator was powered up and connected with the Craterwall electrical grid.

The more dramatic achievement was when Mary and Joan delivered their second children. Mary gave birth to Tommy and Joan had Beverly. The women and their children were healthy. By then, both Norika and Josie were pregnant. The nursery was becoming active with little Martians.

One of the few remaining rocket radios was launched into space above the dust cloud and linked with Marscom and Earth Control. It carried the first news in six weeks back to Earth.

The older children were actively walking. Elizabeth had become quite a climber. They were taken on excursions to the Craterwall farm to become acquainted with the new animals. Those included young Canadian geese,

Mallard ducks, various small birds, southern bull frogs, wild turkeys, Jersey cows, mule deer, chipmunks, squirrels, trout, bass, shrimp, and even some Australian wallabies and koalas, as a tip of the hat to Arnold.

The team was determined to establish substantial biodiversity. New plant species included many berries, grafted apple, peach, cherry, pear, and other small fruit trees. Pond plants were added to provide food for colorful fish.

The team continued their regular morning meetings after breakfast.

Tim said, "This is to ask for your thoughts on whether our experience with the landings should change how we grow our facilities. We have planned on keeping everything conveniently clustered near Craterwall. Now I'm thinking we should spread things out and keep their locations concealed as much as possible. If the terrorist types are able to send explosives to Mars, they could do a lot of damage as things stand now."

Josie said, "One reason for creating our place here was to set an example for peace. I hate to think that could become necessary. Besides, we land our ships with a fairly large landing site ellipse. They probably can't send a ship to Mars—and if they could, they would not be able to hit a target with any certainty."

Rob said, "We all want to continue as before. Still, I agree with Tim. There's nothing wrong with spreading critical projects out a bit. In fact, Arnie and I have been developing an invention that could help us along."

Arnie said, "We've been working on a practical cross-country transport design that we could use in a dust storm. Our concept is for an electric railroad—a combination of surface subway with an underground third-rail power source reached through a slot much like the cable grippers of San Francisco cable cars. That way there would be no obstruction to surface movement by other vehicles and no need for overhead catenaries or towers. We've constructed a scale model to test the concept.

"The railroad would transport smaller freight cars than on Earth as well as pressurized cars for human and other living things. The first line could be run out to the aluminum mine, followed by another line down into the fossa, but using a cog train system to raise the loads of uranium ore up from the fossa floor."

Rob said, "This way, we can move about during another dust storm and even move people in the pressurized cars. This is a relatively lightweight system and the rail and power part is hardly visible. It allows us to begin serious exploration, building safe havens along the way."

Tim said, "I like the exploration prospects for their own benefit and because they allow us to reach out for legitimate reasons. Those will certainly gain support on Earth. It can also allow us to build out without making it obvious that we are also spreading out our facilities. Nearly all of our structures are underground anyway."

Rob had been thinking along the line of the dispersal scenario. They had already begun construction on another structure near Craterwall. They called it Darwin. Construction on a second pebble reactor to support that facility was part of its design. Such a reactor was smaller than most and less likely to risk radioactive exposure.

Ho said, "The pebble reactor is nearing completion. The uranium enrichment systems are promising now that Joan has worked out the kinks. More importantly, these designs are constructed using JPL ideas to improve on our earlier approaches."

They were also well into system testing of three centrifuges for uranium enrichment in the Einstein complex. Centrifuges were the traditional way to stepwise enrich UF_6 from 0.6–0.7 percent U-235 to about 5 percent. The enrichment facility would be paired with a commercial-capacity electrical generation plant needed to power the centrifuges. Once they had stockpiled substantial refined uranium and had the enrichment processing comfortably ahead of production schedule, they would begin to relax.

They were also determined to guarantee redundancy for power for the entire base. One more pebble reactor power plant was in the early stage of construction in the mining and refining area, a facility originally to be named Crater Tall and to be the largest of the new power constructions. It would be renamed Edison once it was in operation.

$$*\qquad*\qquad*$$

Joan had the most recent Mars weather report from Earth. He said, "We finally have a promising outlook for getting out from under the dust. The dust cloud has disappeared entirely in some areas and is thinning to the east of us. Sunlight is beginning to penetrate to the surface. The dark overcast is even turning deep red where the sun is positioned and should spread to a brilliant rose over the entire sky over the next few days."

Mary said, "Let's take our friends out for a ride across the surface as soon as the weather permits."

Everyone was more than excited at the prospect. Finally the first connection was made with the pulse of the Marscom satellite. A waiting message welcomed eight eager humans reestablishing contact with their support team on Earth.

Welcome to the Expanded Mars Team

Welcome back to continuous messaging. We will first be sending the news from Earth just to bring you up to date. We anticipate your many backlogged reports. Everyone here is antsy to get going again. You may need to allow us a little time to get our people back on Mars support schedule.

Your message about having the first pebble reactor online was a great relief. We knew you were at the limit of your power.

We also are very pleased with the arrival of your two new children. Congratulations to Tim and Joan and to Rob and Mary. That made a big splash with the media. We look forward to some pictures.

The media has also been after us for how you are doing. Please be careful how you respond because things here have been changing.

The public is aware and alarmed at the attacks on your landing ships. We suggest that you report technical activity to the media in generalities. Specifics could make you vulnerable. You have a lot of human interest activity that would be appropriate to report.

Earth Control

The first message from Mars was more personal. It was also meant to shore up a positive image with the folks back home after a long absence.

Greetings from Mars

Your daily presence in our lives has been sorely missed. We've been out of interactive contact with you for almost six months. You who support us and news of happenings on Earth provide essential substance in our lives. You are our contact with reality. We may be on another world, but you remain the heart and soul of our lives. Everything we do here is to further the reach of humanity.

We are now Ho, Norika, Arnold, Josie, Rob, Mary, Joan, and Tim plus two small children and two infants. We're all well and very busy. Our living space is much more our home. Arnie and Josie brought mural images of scenes from Earth. We've printed them on our walls, so we share many Earth scenes with you, particularly scenes that are blue and green.

We have grown into our living and work spaces; they now encompass over a quarter of the crater wall. We have forty-one real, intelligent robots (all named) and most of our larger equipment is robotically controlled. We have a large video, music, and reading collection.

We've faced many challenges during our lengthy period under the dust storm. However, we found our facilities were very reliable. We managed to finish all of the corridors and rooms we had excavated, and we made them very livable. The arrival of Arnold, Josie, Ho, and Norika was the highlight of this period and the animals, plants, and equipment they brought have given us many more opportunities.

We deeply appreciate the quick responses from JPL and Earth Control when we were challenged during their landings. We feel very fortunate that matters worked out as well as they did.

We extend our greeting to all who have supported our activities so generously.

Mars Team

Their first activities upon clearing of the dust were to reactivate the mothballed outside equipment with rapid involvement of the many Earth controllers. Those on Earth quickly resumed managing the welcome activities of the additional equipment.

The first new outside project was to begin construction of the railway from Craterwall to Curie at the east end of the Fossa (Fossa Line) where the uranium was being mined. That was quickly followed by construction of the Curie pebble reactor. Meanwhile, rail construction began out to the well and then on to the aluminum mine where its transport supported construction of the pebble reactor.

The Kassei Line was to be their great adventure. Its construction seemed to explode across the landscape. The initial construction was the Well Line from Craterwall, then along Mine Road (Mine Line). Most notably, a new line was being constructed south from the well site to support construction of the Einstein and Fermi facilities and another extended easterly to the space port which was to be kept comfortably distant from the other facilities at the "shore" of the Acidalia Planetia, keeping in mind that arriving rockets were not always trustworthy.

The south-directed construction became known as the Kassei Line as it was intended to eventually reach the considerable distance to the Kassei Valles region. The Kassei Valles was a deep and lengthy gorge extending from near the equator cutting northward across the Tharsis Bulge to eventually turn eastward at about 24 degrees north, emptying into the Chryse Planetia plain near where the 1976 *Viking I* landed. There was a fossae cluster deep in the bulge just east of the north-south segment of the Kassei Valles. The region on the north end of the Kassai included a huge impact crater, a raised mesa apparently carved out by ancient floods, and an extended estuary.

The Kassei Line would run across rough country and had to cross a wide ravine about 90 kilometers south of the well junction.

CHAPTER 33:

AQUIFER ERUPTION

ROB AND ARNIE decided to take a run down the new Kassei Line to the construction area. They wanted to inspect the new construction crossing a huge ravine to confirm what they had seen in photographs. The robots had carved a grade down a gradual slope from the north rim, occasionally using explosives. They had worked their way through the collection of boulders at the foot of the slope and were easily working their way across the floor of the ravine.

Rob noticed that thick layers of ice crystals were embedded in the north wall as they descended through the area once shielded by boulders. The two men followed the construction team across the floor of the ravine and, at night's fall, halted on a high prominence above and in the middle of the ravine's floor. That afforded a better view. They settled in for the evening.

As they relaxed, they could hear small explosions as the robots worked their way up the south slope through the night. They fell asleep to the sound of explosions.

When the explorers awakened, Arnie pointed to explosives being set about midway up the south slope. Rob saw that this would be in a place where the wall was almost vertical and certain to result in a landslide. That would undermine the higher slope as they worked up hill.

The robots had dropped back onto the floor of the ravine. Rob radioed a warning to the robots not to fire the explosives, but it was too late. He immediately raced the hauler toward the peak.

The explosion removed a huge chunk of the south wall, undermining an even larger section, all of which came crashing down in a huge landslide. That was followed by a tremor coming all along the south wall that shook the floor of the ravine. Arnie was watching the action behind them.

"Haul it, haul it! At least a half kilometer section of the south wall is collapsing and seems to be flowing across the ravine!"

Rob saw a huge wall of water tearing across the floor of the ravine. He knew what was happening. He said, "That's the catastrophic collapse of the melted ice embedded deep into the wall. The ice has been melting as we have seen in isolated locations all over Mars. Now it's giving way. I'm hoping this prominence is high and sturdy enough to protect us. Check the air seal on the cabin. If the water crests the prominence, we may be able to float. I'm dumping our equipment attachments. We need to make the hauler as buoyant a possible.

"Control, this is Arnold! The ravine walls are collapsing and a huge wave of water is now racing toward us!"

The control station was always watched—and usually monitored by robots—but Tim was alone on duty. He activated an alarm and immediately began gathering rescue resources.

Rob said, "This is almost certainly due to melted ice from an aquifer along the ravine breaking out. The burden above the aquifer is collapsing into the onrushing water! We can't escape to the north wall. With luck, we'll be high enough on the prominence here in the middle of the ravine."

Arnold said, "Our hauler is airtight—maybe it can float."

Rob said, "This looks like the enormous catastrophic event that produced the Channeled Scablands in Washington! I expect this will scour everything—much like a tidal wave. This is worse than a levee breaking on a huge lake because the burden above the water is collapsing into it. The water is surging around us and washing away its base."

Tim asked, "What can we do?"

Rob yelled, "Just pray. This prominence is fairly large, but there is no telling where the water will stop. It's surging up and down the ravine and up the walls of the prominence, which is only about half as high as the ravine walls."

A mountain of water charged up the wall of the prominence. The crest of the wave showered the hauler and rose high above them. A massive sheet of water crashed down and the frail hauler repelled the liquid, rocking radically and then settled on its rear. The surge descended only to be met by another surge. The prominence shuddered from the enormous impact. The crest of the prominence fell away and the vehicle rocked forward onto its wheels. The dome ceiling shattered and the interior atmosphere burst outward. The electrical system exploded the remaining fuel—and then there was silence.

Tim said, "Rob … answer! Give us anything to work with."

"Mary, take the con. Work the satellite to get a close-up of the site. Joan, we're taking the emergency hauler with the flyer. We'll ride a railcar as far as it goes and then ride the roadbed."

Ho said, "Do you have anything that can float once the waters are calm?"

Tim said, "Right now, the water is evaporating at an enormous rate, which chills the remaining water. I expect it will quickly form a solid ice crust. At this moment, the evaporation is probably creating a cloud of fog that is blanketing the entire water surface. I don't expect you will see much from the satellite until things settle down. We'll have to wait until the ice sheet becomes stable and then run across it to the prominence—if it's still there. We'll use the flyer to recon as soon as we can get it airborne."

Arnold was stunned, but still conscious. Rob was unconscious. Arnold's pressure suit was rapidly leaking air. He popped two emergency patches onto his legs and another on an arm, which still didn't give him a tight suit. He pushed Rob into a pressure bag, sealed it, and then released the reserve of air to pressurize it. He wedged Rob between three rocks. He then positioned his own pressure bag, climbed inside and sealed it, and then released his own atmospheric reserve. Life-support monitors and equipment were built into their pressure suits as well as radios in the bags.

There was an emergency radio beacon on the north side of the ravine. "Control, this is Arnold. Can you read me?"

"Arnie, this is Mary. We're putting together a rescue effort. We're getting some life indications for you, but little from Rob. Report your status."

"Our hauler was blown to shreds from the interior pressure. We're both in patched pressure suits and sealed in pressure bags. We're in a dense fog and presumably still on the prominence. Tell me what to do for Rob. He seems unconscious. I'm picking up weak vital signs from his life-support unit."

"How are you for bleeding and broken bones?"

"I hurt all over. My ribs may be cracked and my left leg hurts. I don't seem to be bleeding seriously. Rob has a welt on the side of his head. I put a pressure pad on his leg where it was bleeding, but otherwise there is nothing obvious. I inserted his IV before I sealed him up."

Mary said, "Crusoe is now monitoring both of you. He'll speak to you as he works."

Tim said, "Actually you seem to be doing fairly well, Arnie. Crusoe is controlling Rob's treatment via radio. Joan and I are heading out down the rail tracks. We'll keep you posted on our progress."

Crusoe had been waiting patiently for his opportunity. He said, "Rob is indeed unconscious, but otherwise he is stable. His life signs are weak. I expect he had a nasty blow to his head. Right now the best we can do is allow the swelling inside his skull some time to go down on its own. He may have a subdural hematoma—a pocket of blood between his brain and his skull. I've started some mild medication through the IV.

"Arnold, you can take some pills to reduce your pain and swelling. I suggest you also insert your own IV in case you fall unconscious."

Ivan had loaded the flyer on the emergency hauler. The vehicle carried a full load of emergency equipment and was essentially an ATV. Joan and Tim climbed aboard and belted in. Ivan rode in a jump seat. There was pallet space for injured people with emergency equipment.

A flatbed railcar was in position for quick placement and lashing of the loader. The railcar was quickly loaded and became part of a small train. Tim guided the train from the hauler as it moved out on the tracks and through the switch that brought it onto the Kassei Line.

Tim brought the train up to maximum safe speed. Joan had been in contact with Craterwall. She said, "There is no significant change in condition. Rob is still unconscious. Arnie is sedated. The emergency reserves in the pressure bags are holding up, but that would be good for only a couple of hours. Their inside temperature is okay. The ravine is still covered with a fog so we can't see anything of the surface.

"We're entering rougher terrain so the track is beginning to meander. The speed of the train has slowed automatically as we approach the ravine. Arnie, can you hear me?"

"I'm right where you left me. I can see a small patch of surface. A thin layer of ice has formed on the surface, but not on our bag, which is warmed. Rob has not moved since we settled in. Whatever pushed us here has gone away."

"Arnie, we're approaching the end of the line near the edge of the ravine. We must be higher than you are. We can see some light through the fog. We're stopping now and will begin unloading."

"Tim, that descent into the ravine was cut out of rough dirt. I have no way of knowing what the water surge may have done to it. Please, do not attempt to descend until you have a flyer's view of the path down into the ravine."

They were so near—yet so far away!

"Mary, can you see a change in the satellite view?"

"Not much so far. The fogbank extends to the east end of the ravine, which is probably as far as the water flowed."

Joan was impatient. She said, "I want to send Ivan to scout the approach to the descent path."

"Okay. Ivan, check the surface conditions to the path. Stop if you sense any softness or a dangerous surface."

"Okay. I'll keep you advised."

They watched as their robot stepped down from the railcar and walked off into the fog. His enhanced sensory equipment allowed him to see and hear better than humans could.

"The way is unimpaired to the top of the path. I'm beginning to go down the path, looking for signs of washout or erosion. I'm using my GPS system for guidance so you will be able to follow my progress and my vision to see for yourselves. I'm picking up debris on the path and a little frost."

Joan said, "Ivan, remember that ice will make a surface slippery. Check your footing before trusting it."

"Understood. I've extended the small spikes on the bottom of my feet. My terrain view still matches the recorded view. I'm at the same elevation as that prominence. I'm picking up the beacons from the emergency balloons. They're about fifty meters lower than here and about 250 meters south by southeast."

"Arnold, Ivan is about three hundred meters above the ravine floor. He's going down."

Everyone was anxious as to whether their route would prove passable. Ivan was continuing with no apparent problems. It appeared that the prominence had blocked the force of the surge. At two hundred feet above the floor, Ivan encountered an ice sheet extending out from the path. He pounded on the ice and found that it was six inches thick when it shattered under the impact. He reached down through the hole, but could not find any ice or water below.

Joan had anticipated this result. She said, "The water reached a maximum level and the surface began to freeze. Meanwhile, the water was extending down the ravine and drained out from beneath the ice. There is no support for the ice so we cannot cross it."

Tim laughed and said, "Can we go underneath or will it collapse?"

"I expect it has already collapsed in spots. We can't cross the top and we cannot go underneath for fear of being buried by tons of ice. No one would have ever expected this kind of experience on Mars."

Tim said, "Okay, we unload the hauler and use it to crush the ice. The hauler will clear away a section of crushed ice and dump it to the side. This could take a while. Ivan, you need to operate the hauler in the manner I just described."

He didn't mention that Ivan would take the risk of an accident.

Ivan climbed aboard the hauler and began down the path until he reached the ice. He gradually advanced until the ice crushed beneath the front wheels for a short distance and then he scooped the crushed ice and dumped it. He dismounted and inspected the result.

"This is what I see. The primary ice sheet is hard, frozen, murky ice—between five and six inches thick. It's suspended, having once floated on the surge water until the surge abated and drained down the ravine. There is a growing coat of rime ice on top of the sheet. I'm working in a fog that's freezing on me and the sheet. My heat is melting the rime ice. I am at the edge of the ice sheet and can see the surface below through the hole I broke. I'm stepping out on the ice sheet now to test its ability to support my weight."

Tim knew that Ivan's weight was much less than normal Earth weight, but his weight was still over a hundred pounds.

"I'm walking on the very edge of the ice sheet and gradually moving out onto the sheet. It's shifting and does not appear to hold my weight."

Joan saw their predicament. She was the lightest person even in her pressure suit.

"Put me in a stretcher frame that will spread my weight. Connect that to the light winch cable. Give me a pair of grippers I can use to pull my frame along the ice. If I break through the ice, just winch me in. Attach a rope alongside the cable. I can pull the rope behind my stretcher. Ivan, you will have to guide me toward Rob and Arnie. I'll anchor the cable to the prominence."

Joan climbed inside the pressurized hauler, took off her weights, and got back into her pressure suit.

Tim knew what daring that took from Joan. He and Ivan made the cable and stretcher frame setup. Joan strapped into the stretcher and began pulling herself out across the ice. It seemed to become more difficult as Joan got farther across the ice—and even more as she tired. They talked by radio.

It seemed like forever. Arnie got out of his pressure suit within the emergency bubble, removed his own weights, got back into his pressure suit, and opened his bubble. Tim guided him down to meet Joan. She was fifty yards out on the ice when the rope gave out. Arnie was on the ice in his pressure suit and very gradually worked his way out to meet her. She threw a coiled rope as far as she could and Arnie finally reached it. He reversed direction by pulling the rope tied to his pressure suit and eventually reached the shore. Joan gripped the rope tightly and Arnie hauled her in.

Mary brought another hauler by train out to the ravine just in time to hear of their success, but there was no cause for celebration. Rob was still unconscious. Joan and Arnie pulled the stretcher up to where Rob was wedged. They loaded him onto the stretcher and tugged him down to the shore and onto the ice. Joan and Tim winched the stretcher across the ice and to the shore. None of this went quickly. Ivan and Tim unloaded Rob and placed him into Mary's heated hauler, which was taken immediately by train to Craterwall where they had better emergency equipment.

Arnie and Joan pulled the stretcher across the ice and back to the prominence by tugging the rope. Since Arnie was injured, he was winched and loaded into the other hauler next. Finally, Joan retrieved the stretcher and climbed aboard so the winch cable could also pull her to safety. She was totally exhausted. Tim and Ivan disconnected the cable and rope and lifted the stretcher into the hauler's second stretcher bay. Ivan took charge of the train to return everyone to Craterwall. Tim stayed in the back of the hauler with Joan and Arnie. The last group pulled into the dock at Craterwall more than an hour after dark.

Everyone had gathered in the infirmary. Rob, Arnie, and Joan were the worst. The two men were placed on mandatory bed rest with medication, particularly for Rob who was still semiconscious.

Joan was up in a matter of days. The augmented team had already survived three emergencies together. The new team members had already been initiated by fire. They were moving rapidly toward the strong bond of comradeship and confidence that they would need in the days, months, and years ahead.

Chapter 34:

Reflection

Tim had been reflecting upon their experiences as they recovered. After three weeks , they gathered for their normal morning planning session.

Tim said, "We've really come together, but one element may be weak. Our original four are more likely to be critical of each other's ideas. We've always enjoyed that give and take. In fact, that may be what saw us through our first year. We may have enjoyed an advantage because of those experiences, but we are all a single group now and we really want everyone to jump in with any idea that could help us find a better approach."

Arnold said, "Do you have any idea how highly you are regarded back on Earth?"

Mary lightly shook her head. "We did sense that there was much more there when we were fighting to bring you four into our team. We were pleasantly surprised when the president backed us up with the G-10. In fact, we ended up really scrambling to respond quickly to his shift in position."

Joan said, "We jumped upon the opportunity as a group. Are you telling us that it was more than an executive decision?"

Josie said, "You really don't know what was behind it? The people on Earth and particularly those on the NASA-JPL Mars team think you could walk on water—if there was any on the surface here. Come to think of it, we pretty much did slide on ice."

Norika said, "We've been slow to add to your thoughts because we still hold you all very much in awe. You reinforced that by saving us from our crash landings. When we came inside and saw the scope of this base and the extent of your menagerie and variety of agriculture and plant life, it blew us away. We knew it was here, but we had seen only snippets. You had created far more than we expected.

"There is one other matter we have not brought up. We were taken into a hush-hush meeting with Gus and Dave and some people who have been monitoring your health. They're convinced that you're different from when you first arrived. Frankly, we all feel the same way. In the beginning, you were following a plan and struggling to get things together, but in the two years since, you're taking much more initiative and you're now ahead of the NASA and JPL's brain trust. We were really following you entirely, but we're now catching up."

Joan smiled and said, "Yes, you are. But that is to be expected. It's called experience."

Arnold shook his head and said, "We think it's more than that. We think it's environmental. We're all experiencing low gravity, but zero gravity is too little. Our hearts are able to circulate our blood to our brains and our entire bodies more easily. But it's more than that. Our oxygen level is a little higher for each breath. It's more than mental acuity; it's your ability to cogitate. You have no diseases and neither do the children. Furthermore, the children are developmentally more advanced than their age group on Earth and they're very healthy. We're eating a diet that's different and we're all exercising rather strenuously. As a result, we are in virtually every category better than our peers back home. You, in particular, are noticeably superhuman and I expect aging more slowly."

Tim did not take this development as all positive. He said, "This is all subjective. I'm concerned that if this becomes the general view, then we could become a curiosity and feared rather than respected."

Rob said, "But we are all one now. We depend upon one another. You've given us the opportunities we had hoped to have. You're well on the way toward enriching uranium and we're constructing electrical generators that will be able to produce much more power than we have now. Power is everything here. You underestimate our need for your contribution. You had already proven yourselves, but our experience together in escaping the flood made us a family. In your way, you've made yourselves worthy of our awe."

Tim said, "We need to think ahead. If we're to lead the way into this outreach, then we must consider the potential. We've lived only in a minute space on this world. We have considered the possibilities and we should evaluate them now."

Tim said, "Joan, Rob, Mary, and I need to tell you something about our secret personal expertise. Collectively we call this 'stardust'. Some of this is already evident because of our discovery of rich uranium ore at our doorstep. Gus did not tell us that each of us had this expertise. We revealed it among ourselves when we were pushed into the uranium enrichment business. Enrichment was Joan's contribution. It allows us to generate all the local power we might dream of for our habitat, our mining and industry, and transportation across Mars. These talents give us opportunities that we need to discuss.

"For example, I know how to build a proven plasma rocket engine. We can build a rocket ship that can ferry people and cargo between Mars orbit and Earth orbit much more economically and in one third of the time that it took to get here. Since we already can build chemical rockets for launches and landings, together they put us ahead in the space transportation business.

"Rob has worked with nuclear breeder reactors that produce plutonium from non radioactive uranium 238. That is tricky, but it puts us at the forefront of nuclear power generation.

"Mary knows the latest in medical processes employing in vitro fertilization and artificial uterus growth of animal embryos ending in the birth of natural offspring. We have frozen donated eggs and sperm from exceptional human donors. That means we can populate Mars as we may wish.

"As I said, 'stardust.' Collectively we must decide how and when to employ this knowledge. We kept this technology to ourselves until just a few months ago. Welcome to the inner circle."

Arnold said, "Gus literally loaded this mission as a survival project—a Noah's Ark to save life on Earth and send it to Mars where you were building habitats. If somehow devastation happened across Earth, at least that much could be saved."

Rob said, "Actually we're a demonstration for them. We're expected to show them how they can survive. I personally think that it's wishful thinking."

Joan said, "With all of that taken into account, this project from the outset needs abundant energy, a large robotic work force, a large viable habitat, and a substantial educated human nucleus. I expect we can realize some of his goals. Our present group of people and the time needed to grow and educate people would seem to make that prospect difficult to achieve. I expect we'll just use our new talents as judiciously as possible until we have a revelation."

*   *   *

Gus was reflecting as well. His dream of a human colony on Mars was now very real. They had all they needed to thoroughly explore Mars—and even to reach the entire family of planets.

People saw not just the advantage of the Mars colony, but with plasma rockets and an outpost on a distant planet, they were truly able to go where no person had gone before. More appropriately, mankind could extend its reach and grasp farther across the heavens and seek to become one with the universe.

REFERENCES

Ares I with *Orion* Crew Exploration Vehicle

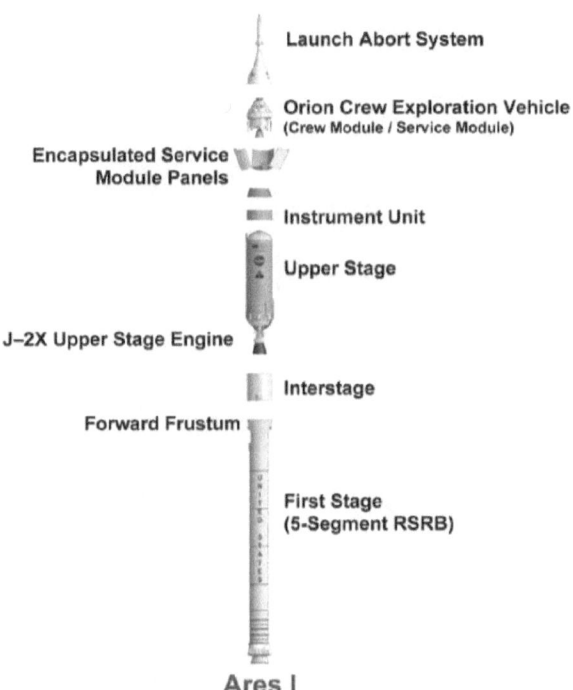

Launch Abort System

Orion Crew Exploration Vehicle
(Crew Module / Service Module)

Encapsulated Service
Module Panels

Instrument Unit

Upper Stage

J–2X Upper Stage Engine

Interstage

Forward Frustum

First Stage
(5-Segment RSRB)

Ares I

National Aeronautics and Space Administration
www.nasa.gov/centers/marshall
www.nasa.gov

NASA'S Ares I-X Rocket
Completes Successful Flight Test

NASA's *Ares 1-X* test rocket lifted off at 11:30 a.m. EDT Wednesday, October 28, 2009, from NASA's Kennedy Space Center in Florida for a two-minute powered flight. The test flight lasted about six minutes from its launch from the newly modified Launch Complex 30B until splashdown of the rocket's booster stage nearly 350 miles down range.

"This is a huge step forward for NASA's exploration goals," said Doug Cooke, associate administrator for the Exploration Systems Mission Directorate at NASA Headquarters in Washington. "*Ares 1-X* provides NASA with an enormous amount of data that will be used to improve the design and safety of the next generation of American spaceflight vehicles—vehicles that could again take humans beyond low Earth orbit."

The 327-foot tall *Ares 1-X* test vehicle produced 2.6 million pounds of thrust to accelerate the rocket to nearly 3 g's and Mach 4.76, just shy of hypersonic speed. It capped its easterly flight at a sub-orbital altitude of 150,000 feet after the separation of the first stage, a four-segment solid rocket booster.

Parachutes deployed for recovery of the booster and the solid rocket motor will be recovered at sea for later inspection. The simulated upper-stage Orion crew module and launch abort system will not be recovered.

"The most valuable learning is through experience and observation," said Bob Ess, *Ares 1-X* mission manager. "Tests such as this—from paper to flight—are vital in gaining a deeper understanding of the vehicle from design to development."

Wednesday's flight offered an early opportunity to test and prove hardware facilities and ground operations—important data for future space vehicles. During the flight, a range of performance data was relayed to the ground and also stored in the onboard flight data recorder. The seven hundred sensors mounted on the vehicle provide flight test engineering data to correlate with computer models and analysis. The rocket's sensors gathered information in several areas, including assembly and launch operations, separation of the vehicle's first and second stages, controllability and aerodynamics, the reentry and recovery of the first stage and new vehicle design techniques.

Source:
http://www.nasa.gov/home/hqnews/2009/oct/HQ_09-252_Ares_I-X_Success.html

Schematic Base Design within an Impact Crater Wall

The base was planned to be the primary facility where most operations would be housed. The space occupied by people would be the smallest part of the base. The larger parts would be for industrial facilities, equipment storage and maintenance, and for agriculture and animal husbandry. Utilities would be distributed throughout.

SCHEMATIC OF CRATERWALL

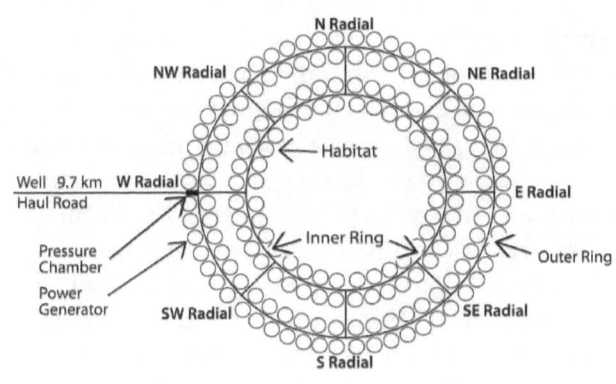

Side View

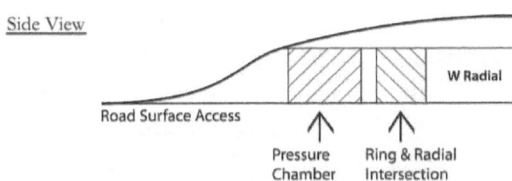

Front View

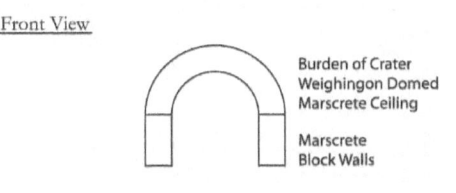

Schematic drawn by Tim Joseph

The thick earthen wall of the Genesis Crater was to become the base. The crater was barely a kilometer across and thus a little over three kilometers around. It was named simply Craterwall. A geometrical design would be followed in laying out the corridors of the base. Since the crater was shaped like a doughnut or bagel, a circular notation would be used to designate the corridor passageways and address the adjoining cells or rooms. There would be three concentric circular corridors and eight radial corridors at the primary points of a topographic compass (true north). Mars has no magnetic field so a real magnetic compass would not function.

The radials were like spokes on a wheel with NW for the northwestern corridor extending from the inner edge of the crater at the central depression to the outer edge of the crater's wall where the industrial facilities would be installed.

The three circular corridors would be designated Outer, Middle and Inner. Outer and Inner corridors would be for pedestrian use and light transport and provide access to the rooms. Middle corridor was for heavy transport. The addresses of the rooms or cells would be assigned numerically, progressing north and south from the East-West corridor and would be odd on the outside and even on the inside of each circular corridor much like odd houses are on one side of a street and even on the other side.

The rooms along the corridors were intended to be all of the same size and design with a short corridor access from the ring corridor. Each room would be half a spherical dome raised two meters from the floor. The spherical diameter was eight meters. The crown of the domed ceiling would be six meters, over eighteen feet high. The lower two meters would be compacted tight and fused outward from the wall as a buttress into the natural crater, both to hold the ceiling and to prevent air pressure from expanding the base of the room.

Storage space for vehicles and larger equipment was to be provided near the entrances on the radial corridors. The depression inside the crater was to be eventually hollowed out and domed like a honeycomb. It would become a large ice house for frozen blocks of pure ice, their reservoir. The domes would be covered with soil for ballast.

Construction of the individual rooms would begin with a small tunnel around the perimeter of the planned room. This approach would allow construction of the buttressed lower section of the dome with Marscrete blocks on the inside. The work tunnel would then be enlarged upward allowing the dome to grow, but maintaining the natural pillar of earthen material that was supporting the center of the growing dome. The dome would arch inward and all of the way around its circumference as it grew, thus providing more and more support for the weight of the increasing crater wall load.

Eventually, the top of the dome would be reached and the supporting column cut loose from the ceiling. As a final act, backfill would be packed into the remaining space behind the crown of the dome, the last stone set in place and the Marscrete backfill fired.

The air pressure in spaceships was maintained at half Earth normal. That was a problem for the base since the growing animals required substantial air pressure for them to develop normal lung size. People and animals would wear weights to encourage development of muscles. When the base was filled with air at 10 psi, the vertical upward pressure would be a total of about 1,000 tons per room. That pressure and more would be countered with the crater's ballast weight, which is the weight under Mars gravity of the crater wall material directly above the chamber.

Each room would have fifty square meters of floor space (about five hundred square feet). They would be spaced eight meters apart; thus there would be space for about 150 rooms / chambers on each side of each corridor or space for about 600 rooms along the Inner and Outer Ring hallways of the crater, allowing space for the radial corridors and the like.

The Thor project needed the robotic equipment to construct at least sixteen of these rooms in the first year to accommodate the industrial activities, the power plant, the minimum agricultural projects, utility functions, and domicile space for the incoming astro-colonists. There were also segments of the circular corridors and the essential western radial to be built.

BIOGRAPHY

Bert Tucker holds a bachelor's degree in engineering from West Point (1956) and a master's degree plus in physics and mathematics from Louisiana State University (1964), including an experimental and theoretical thesis on the fluid dynamics of superfluid liquid helium under a grant from NASA. He was a lieutenant and captain in the Corps of Engineers, engineer company commander (Germany, 1960), airfield operations officer (Fort Polk, 1961–62), and served as an airplane pilot, helicopter pilot, and helicopter instructor pilot. He was an FAA-qualified commercial pilot and was qualified as a military parachutist.

He participated in a course presented by senior NASA instructors on the design of early spacecraft and implementation of the then new Apollo project while he was a graduate student at LSU. No missions have carried men beyond Earth orbit since the Apollo project sent the first men to the moon. For further background, see the acknowledgments.

He worked with and consulted to Wall Street firms for more than a quarter century. He developed and managed the data quality of financial market data systems that employed early global satellite communication systems, acquiring data from one hundred exchanges around the world. He pioneered early derivative securities reporting methodologies. He developed systems to calculate many instantaneous, complex market data indices. He developed encryption methods to secure proprietary market data.

He is a member of the West Point Society of the Mid Hudson Region and was the chair of a series of seven West Point conferences on leadership and ethics development for high school students sponsored by the WPSMHR. He is a member of the Wayne, New Jersey Rotary Club and has served as chair for Rotary District 7490 for ten leadership and ethics conferences sponsored by numerous Rotary Clubs. The seventeen conferences were presented by West Point officers and cadets who were leaders in the West Point Cadet Honor and Respect programs.

www.ingramcontent.com/pod-product-compliance
Lightning Source LLC
Chambersburg PA
CBHW031943170526
45157CB00002B/374